KT-244-359

Hume

Past Masters

A. J. Ayer

HUME

Oxford   Melbourne   Toronto
OXFORD UNIVERSITY PRESS
1980

Oxford University Press, Walton Street, Oxford OX2 6DP

OXFORD LONDON GLASGOW
NEW YORK TORONTO MELBOURNE WELLINGTON
KUALA LUMPUR SINGAPORE JAKARTA HONG KONG TOKYO
DELHI BOMBAY CALCUTTA MADRAS KARACHI
NAIROBI DAR ES SALAAM CAPE TOWN

First published as an Oxford University Press paperback
1980 and simultaneously in a hardback edition

British Library Cataloguing in Publication Data

Ayer, Sir Alfred Jules
Hume. – (Past Masters).
1. Hume, David
I. Series
192   B1498   79-41293

ISBN 0-19-287529-9
ISBN 0-19-287528-0 Pbk

Printed in Great Britain by
Cox & Wyman Ltd, Reading

Preface

With the exception of the short biographical chapter, for which I am greatly indebted to Professor Ernest C. Mossner's excellent book *The Life of David Hume*, this book reproduces the text of the four Gilbert Ryle lectures which I delivered at Trent University, Ontario, in March 1979. I was all the more pleased to be invited to give these lectures, as Gilbert Ryle was my own tutor in philosophy, and I wish to express my gratitude not only to the sponsors of the lectures, the Machette Foundation and the Victoria and Grey Trust Company, but also to the members of the Department of Philosophy and many of their colleagues at Trent for the warmth of the hospitality which they showed to me.

In quoting from Hume's philosophical works, I have made use of the following texts, of which the first three are available in paperback.

A Treatise of Human Nature, edited by L. A. Selby-Bigge; second edition revised by P. H. Nidditch; including Hume's *Appendices* and *Abstract*. Oxford University Press, 1978.

Enquiries concerning Human Understanding and concerning the Principles of Morals, edited by L. A. Selby-Bigge; third edition revised by P. H. Nidditch. Oxford University Press, 1975.

Dialogues Concerning Natural Religion, edited with an introduction by Norman Kemp Smith, including Hume's *My Own Life* as a supplement. Bobbs-Merrill, 1977.

Essays Moral Political and Literary, Vol. II, edited by T. H. Green and T. H. Grose. Longmans, 1875.

At the wish of the Oxford University Press, I have put my references to these works within brackets in the text, identifying them by the letters T, E, D, and G respectively. The numerals following the letters refer to pages. My only other references of this sort are to pages of Mossner's book, identified by the letter M.

For all his other literary achievements, including his celebrated *History of England*, Hume was first and foremost a philosopher, and apart from the opening chapter, in which a sketch of his life is given, this book is entirely devoted to an exposition of his philosophy.

I owe thanks to Dr Henry Hardy of the Oxford University Press for commissioning this book from me, and once again to Mrs Guida Crowley for typing my manuscript and helping me to correct the proofs.

A. J. Ayer
10 Regent's Park Terrace, N.W.1.
18 April 1979

Contents

To Raymond Klibansky

1 Life and character

David Hume, to my mind the greatest of all British philosophers, was born at Edinburgh on what, in the old calendar, was 26 April 1711. In his valedictory *My Own Life*, an autobiography running only to five pages, which Hume composed in April 1776, four months before his death, he showed pride in coming of good family, on both sides. His father, Joseph Home, combined the profession of law with the ownership of an estate at Ninewells in Berwickshire, which had belonged to the family since the sixteenth century, the family being, as Hume put it, 'a branch of the Earl of Home's or Hume's' (D 233), which was in our days to produce a Conservative Prime Minister; his mother, Katherine, was 'the daughter of Sir David Falconer, President of the College of Justice', and one of her brothers inherited a peerage. The couple had three children, of whom David was the youngest, his brother John being born in 1709 and his sister Katherine a year later.

Joseph Home died in 1713, while David was still an infant. The estate passed to the elder son and David was left with a patrimony of some £50 a year, which even in those days was not quite enough to make him financially independent. It was planned that he should follow his father's example and become a lawyer. Their mother, who did not remarry, managed the estate until John was old enough to take charge of it. By all accounts, David was devoted to her, as well as to his brother and sister. She was an ardent Calvinist, and brought her children up in the faith, which David rejected, together with all other forms of Christianity, in his teens. That this did not impair his relations with his mother suggests that he concealed it from her or at least did not obtrude it. Throughout his life he was of a peaceable disposition and averse from engaging in public as well as private controversy, though not at all lacking in the courage of his own convictions, however unorthodox, or reluctant to express them in print. The story of his mother's saying 'Our Davie's a fine good-natured crater, but uncommon wake-minded' is not supported by any documentary evidence. If it is true, it may be the expression of a feeling of exasperation at the time which it took him to become financially independent of the family estate.

In 1723, when David Hume was not quite twelve years old, he went with his elder brother to the University of Edinburgh. They were there for the best part of three years and left, as was quite common in those times, without taking a degree. The arts course, in which they were enrolled, comprised Greek, logic, metaphysics and natural philosophy, now better known as physics, as compulsory subjects. There were also elective courses in other subjects such as ethics and mathematics. The level of the lectures seems to have been fairly elementary, but it is probable that Hume gained some knowledge at this stage of the seminal work of Isaac Newton and John Locke. All that he himself says about his university studies is that he 'passed through the ordinary Course of Education with Success'.

Having returned to Ninewells, Hume tried to settle down to the study of law, but very soon gave up the attempt. The passion for literature, understood as including history and philosophy, to which he referred in his autobiography as 'the ruling Passion of my Life and the Great Source of my Enjoyments', proved too strong, to the point where, in his own words, he 'found an insurmountable Aversion to anything but the pursuits of Philosophy and General Learning' (D 233). Though he speaks of Cicero and Virgil as the authors whom he was 'secretly devouring', in place of the jurists whose works his family believed him to be studying, his thoughts ran chiefly on philosophy, and it was in 1729, when he was still only eighteen years of age, that there opened up to him 'the new scene of thought' which was to be displayed in his first and eventually most famous book, *A Treatise of Human Nature*.

The excitement of this discovery and the intensity with which he worked upon it combined to impair Hume's health. His disorder was psychosomatic and a course of regular physical exercise, supported by an ample diet, turned him within two years from a 'tall, lean and rawbon'd' youth into what he describes as the 'most sturdy, robust, healthful-like Fellow you have seen with a ruddy Complexion and a cheerful Countenance'. Nevertheless he remained subject to attacks of nervous depression, with physical symptoms such as palpitations of the heart, and the local physicians, to whom he frequently resorted, were unable to cure him. Eventually, he himself decided that he had better give up his studies, at least for the time being, in order to 'lead a more active life', and in February 1734 he left Scotland for Bristol, where he had been offered a post as clerk in a firm of sugar-merchants. His decision may have been influenced by the fact that he was shortly to be cited by a local servant-girl before an ecclesiastical court, presided over by his uncle, as the father of her illegitimate child. The charge was not considered proved against him and did not, even locally, damage his reputation. There is, indeed, later evidence that he remained susceptible to women, though he never married and was of too calm a temper, and too thoroughly immersed in intellectual pursuits, to qualify as an amorist.

Though he made some good friends in Bristol, it took Hume no more than four months to decide that the life of commerce did not suit him. It has been suggested that he was dismissed from his employment because of his insistence on criticising his employer's literary style (M 90). Whether this is true or not, there is no doubt that Hume was happy to be left free to concentrate on his philosophy. The most lasting result of his stay in Bristol was that he there changed the spelling of his name from Home to Hume, to accord with its pronunciation.

Having resolved to devote himself to the writing of his *Treatise*, Hume migrated to France, probably on the ground that he could manage better there on his small private income. After a short stay in Paris, where he obtained some useful introductions from a fellow Scotsman, the Chevalier Ramsay, he spent a year at Rheims and two years at the small town of La Flèche in Anjou, the site of the Jesuit College where Descartes had been educated. He made friends among the Jesuit Fathers and took advantage of their extensive library. By the autumn of 1737 the greater part of the book was written, and Hume returned to London to find a publisher for it.

This did not prove so easy as he had hoped. It was a year before he succeeded in making a contract with John Noon for an edition of a thousand copies of the first two 'books', entitled 'Of the Understanding' and 'Of the Passions', for which he received

£50 and twelve bound copies. The work was published, anonymously, at a price of ten shillings, in January 1739, under the general title of *A Treatise of Human Nature: Being an Attempt to introduce the experimental Method of Reasoning into Moral Subjects*. The third 'book', 'Of Morals', was not yet ready for publication. Its appearance was delayed until November 1740, when it was published, this time by Mark Longman, at a price of four shillings.

The reception of the *Treatise* was a great disappointment to Hume. In his own words, 'Never Literary Attempt was more unfortunate than my Treatise. It fell *dead-born from the Press*, without reaching such distinction as even to excite a murmur among the zealots (D 234). This is not altogether accurate. It is true that not all the copies of Noon's edition were sold in Hume's lifetime, but the work was noticed in English and foreign journals, and obtained three reviews of considerable length. The trouble was that the tone of the reviews was predominantly hostile, and on occasion disdainful. Hume believed that the hostility arose largely from a misunderstanding of his views, and he sought to remedy this by publishing in 1740 an anonymous sixpenny pamphlet, advertised as *An Abstract of a late Philosophical Performance, entitled* A TREATISE OF HUMAN NATURE, *&c. Wherein the chief Argument and Design of that Book, which has met with such Opposition, and been represented in so terrifying a Light, is further illustrated and explain'd*, but appearing under the less aggressive title of *An Abstract of a Book lately Published. Entituled, A Treatise of Human Nature, &c. Wherein The Chief Argument of that Book is further Illustrated and Explained.* This pamphlet fell into oblivion until a copy of it was discovered and identified in the late 1930s by Maynard Keynes, and published with an introduction by himself and Piero Sraffa under the title *An Abstract of a Treatise of Human Nature, 1740: A Pamphlet hitherto unknown by David Hume*. The abstract draws particular attention to Hume's theory of causation, which was, indeed, the feature of the *Treatise* for which it was later to become most celebrated.

Hume came to think himself largely responsible for the failure of the *Treatise*, because of its defects of presentation, and was later disposed to disown it. The first sign of this is to be found in the preface to the first of the two volumes of *Essays, Moral and Political*, published respectively in 1741 and 1742, in which, remaining anonymous, he is described as a 'new Author'. The essays, which were brought out by Andrew Kincaid in Edinburgh and numbered twenty-seven in all, were of varying degrees of seriousness, and covered a large range of topics, including criticism, manners, philosophy and politics. They were favourably received, especially the political essays on such subjects as 'The Liberty of the Press' and 'The first Principles of Government'. One that aroused particular interest was 'A character of Sir Robert Walpole', a harsher appraisal than Hume wanted to sustain when that statesman had fallen from power. For this reason, no doubt, he did not reprint the essay in later editions of the work. He also omitted several of the lighter pieces with such titles as 'Of Love and Marriage' and 'Of Impudence and Modesty'.

The publication of these essays not only made Hume some money, amounting perhaps to £200, but emboldened him to become a candidate for the Professorship of Ethics and Pneumatical Philosophy at Edinburgh University. The suggestion that he should apply was made to him in 1744 by his friend, John Coutts, the Lord Provost of Edinburgh. The holder of the Chair, Alexander Pringle, had been on leave for the past two years, serving abroad as an army doctor, and his appointment as Physician-General to the Forces in Flanders did not seem compatible with his remaining a professor in Edinburgh. There was then no overt opposition in the Town Council to the choice of Hume as his successor. Unfortunately, however, Pringle delayed his resignation till Coutts had ceased to be Lord Provost and the zealots, whom Hume had after all offended, had time to gather their forces. A pamphlet, entitled *A Letter from a Gentleman to his friend in Edinburgh,* which Hume published anonymously in 1745, denying that he had rejected, as opposed to explicating, the proposition 'that whatever begins to exist must have a cause', or that the argument of his *Treatise* led in any other way to atheism, failed to appease them. The Chair was offered in the same year to Hume's friend and mentor, Francis Hutcheson, who was Professor of Moral Philosophy at the University of Glasgow, and when Hutcheson declined it, the Council chose to promote the lecturer who had been doing Pringle's work.

Still lacking the financial security which the appointment would have given him, Hume accepted an offer of a salary of £300 a year to act as tutor to the Marquis of Annandale, an eccentric young nobleman, soon to be declared insane, who lived near St Alban's at a convenient distance from London. In spite of his employer's vagaries, and the ill-will shown him by an influential member of the family, Hume was sufficiently contented with his position to be willing to consider retaining it at a lower salary. No doubt the reason was that it allowed him leisure to write. It was at this time that he began work on his *Philosophical Essays Concerning Human Understanding*, later to be entitled *An Enquiry concerning Human Understanding*, which was designed to supersede the first book of the *Treatise*, and most probably also wrote his *Three Essays, Moral and Political*. Both works were published in 1748.

The *Enquiry* is, indeed, a much better written work than the *Treatise*, from which it differs more in emphasis than in argument. The central issue of causality is brought more into the foreground, and it is less encumbered with what would now be reckoned as psychology. There are also sections of the *Treatise*, such as that on Space and Time, for which it has no counterpart. On the other hand, it includes a chapter 'Of Miracles', which Hume had omitted from the *Treatise* out of prudence. The central argument of this chapter 'That no testimony is sufficient to establish a miracle, unless the testimony be of such kind that its falsehood would be more miraculous than the fact, which it endeavours to establish' (E 115–16), with its iconoclastic implications, procured Hume more fame among his contemporaries than anything else in his purely philosophical work.

Three Essays, Moral and Political, which appeared in February 1748, was the first of Hume's books to which he put his own name, a practice he was thenceforward to continue. The essays were prompted by the rebellion of the Young Pretender, and Hume said of them, before their publication, that 'One is against the original Contract, the system of the Whigs, another against passive Obedience, the system of the Tories: A third upon the Protestant Succession, where I suppose a Man to deliberate, before the Establishment of that Succession, which Family he should adhere to, and to weigh the Advantages and Disadvantages of each.' In fact the essay on the Protestant Succession was not published till 1752, and was replaced in the 1748 volume by an essay on 'National Characters'. Hume was in no degree a Jacobite, but he wrote a pamphlet in defence of his friend Lord Provost Stewart, who was arraigned in 1747 for having surrendered Edinburgh to the rebels, though owing to the timidity of the printers the pamphlet was published only after Stewart's acquittal.

Hume's willingness to compromise over his tutorship availed him nothing, for in April 1746 he was dismissed, with a quarter's salary owing to him which he may, some fifteen years later, have succeeded in getting paid. He thought of returning to Scotland, to a home bereft of his mother, who to his great sorrow had died the year before, but was prevented by an offer from a distant kinsman, General St Clair, to act as his Secretary on an expedition which the General had been appointed to lead to Canada, with the intention of helping the English colonists to expel the French. While the expedition was waiting at Portsmouth for a favourable wind, Hume was promoted from his Secretaryship to be Judge-Advocate of all the forces under St Clair's command. The wind never did become favourable, and the expedition was diverted to Brittany, where it failed to take the town of L'Orient, abandoning the siege just at the moment when the French were deciding to surrender, and returned to England without having accomplished anything of note. General St Clair appears to have been more unlucky than culpable, and his conduct of the expedition was later to be defended by Hume in print against the ridicule of Voltaire. Once more Hume had to wait many years before he was able to extract from the Government the pay due to him as Judge-Advocate.

After the expedition was disbanded, Hume returned briefly to Ninewells, but early in 1747 he was back in London, having accepted an invitation from the General to serve as one of his Aides-de-camp 'in his military Embassy to the Court of Vienna and Turin'. He wore the uniform of an officer, which probably did not become him. According to an irreverent young witness 'the Corpulence of his whole Person was far better fitted to communicate the Idea of a Turtle-eating Alderman than of a refined Philosopher' (M 213–14). The same observer, though subsequently proud of his acquaintance with Hume, commented on the disparity between his mental powers and the vacancy of his countenance and sneered at his retention of 'the broadest and most vulgar Scottish accent' in speaking either English or French.

Hume remained at Turin until the end of 1748, so that he was absent from England during the period of the publication of the *Three Essays,* the first volume of the *Enquiry,* and a reissue of *Essays, Moral and Political,* which laid the foundation of his literary reputation. The great French writer Montesquieu was so impressed with these *Essays* that he sent Hume a copy of his *L'Esprit des loix,* and the two men corresponded regularly for the remaining seven years of Montesquieu's life.

Hume himself was slow to realise that the tide was turning in his favour, at least if we are to believe his autobiography, in which he speaks of his mortification at finding on his return to England that neither the *Enquiry* nor the reissue of his *Essays* had achieved any great success. This was not however a discouragement but rather a stimulus to his literary ambition. Returning to Ninewells, he completed by 1751 the *Enquiry concerning the Principles of Morals,* designed to replace Book III of the *Treatise,* and considered by Hume 'of all my writings historical, philosophical or literary incomparably the best' (D 236). In the following year he published his *Political Discourses,* and during this period he also began work on his *Dialogues concerning Natural Religion* and engaged in research for his *History of England.* At the same time his work began to attract criticism. In his own words, 'Answers by Reverends and Right Reverends came out two or three a year' (D 235), but Hume maintained his fixed resolution 'never to reply to anybody'.

This hostility did not extend for the most part to the *Political Discourses,* though they did not escape being placed on the Roman Catholic Index, in 1761, along with all Hume's other works. The *Discourses,* described by Hume as 'the only work of mine that was successful on its first Publication', were originally twelve in number, of which only four were strictly political. One was concerned with the relative populousness of the ancient and modern worlds and the other seven were contributions to what is now called economics. Hume was a strong advocate of free trade, and his essays in some degree foreshadowed the theory developed by his young friend Adam Smith in his celebrated book *The Wealth of Nations*, the first volume of which was read with admiration by Hume a few months before his death.

In 1751, John Home married, and David and his sister set up house in Edinburgh, moving to slightly more luxurious quarters as his fortunes improved. Apart from his literary earnings, his appointments at Vienna and Turin had left him 'master of near a thousand Pound' and his sister had a small private income of £30 to add to his £50. Though he speaks of his frugality, he seems to have led an active social life, being frequently entertained by his numerous circle of friends, including many of the moderate clergy, and returning their hospitality. He was, however, prepared to move to Glasgow, if he had been able to secure the University Chair of Logic which Adam Smith vacated in 1752 to succeed to the Chair of Moral Philosophy, but though he had the support of other professors besides Adam Smith, the opposition of the zealots again prevented his appointment.

Hume was in some degree consoled for this failure by being made Librarian to the Faculty of Advocates at Edinburgh. The salary was only £40 a year and Hume refused to take it after 1754, when the Curators rejected, on the ground that they were indecent, three books that he had ordered, one of them being the *Contes* of La Fontaine. Hume did not resign till 1757, but compromised in the meantime by giving the money to his friend Blacklock, the blind poet. The advantage to Hume of the position was that, the library being exceptionally well-stocked, it gave him access to the books that he needed for the writing of his history. He appears still to have had access to the library after he had resigned its Keepership in favour of his friend the philosopher Adam Ferguson.

The six volumes of Hume's history appeared in an unusual order. It began with the Stuarts, the first volume, covering the reigns of James I and Charles I, and the second, which continued the story till the fall of James II, being published respectively in 1754 and 1756. The next two volumes, which came out in 1759, were devoted to the Tudors, and the work was completed by the publication in 1762 of two volumes spanning the centuries between the invasion of Julius Caesar and the accession of Henry

VII. The first volume was a failure at the outset, partly because its attempt to be fair to both sides in the conflict between King and Parliament irritated the Whigs without satisfying the Tories, and partly, it would seem, because of a conspiracy into which the London booksellers entered against the Edinburgh firm to which Hume had entrusted it. Eventually this firm found it advisable to transfer its rights to Andrew Millar, Hume's usual publisher, who then brought out the subsequent volumes. These were much more successful, both critically and financially. The sums for which Hume sold the rights of the various volumes amounted in all to more than £3000, and the work came to be considered by Hume's contemporaries as an outstanding achievement, to the point where he was esteemed even more highly as a historian than as a philosopher. Thus Voltaire went so far as to say that 'nothing can be added to the fame of this *History*, perhaps the best ever written in any language' (M 318). A much later verdict of Lytton Strachey's, in an essay on Hume included in his *Portraits in Miniature*, that Hume's book, 'brilliant and weighty as it was, must be classed rather as a philosophical survey than a historical relation', comes nearer the mark, but the *History* remains very well worth reading, if only for its wit and the beauty of Hume's style.

In the course of publishing his *History* Hume brought out in 1757 another volume of essays entitled *Four Dissertations*. The most important of them was 'The Natural History of Religion'. The second, 'Of the Passions', was a condensation and revision of the second book of the *Treatise*. The third and fourth were 'Of Tragedy' and 'The Standard of Taste'. The essay on the standard of taste took the place of an essay on Geometry and Natural Philosophy which Hume was dissuaded from printing by his friend Lord Stanhope, who was a mathematician. Having discarded the mathematical essay, Hume planned to bring the number of Dissertations up to five by adding to the first three mentioned an essay 'Of Suicide' and one called 'Of the Immortality of the Soul', but his publisher, Millar, was afraid of the consequences of their being taken as a further affront to religion, and Hume withdrew them. Copies of the manuscripts were in private circulation, but the essays were never included in any authorised edition of Hume's works, though unauthorised versions appeared in 1777 and 1783, and they are to be found among 'Unpublished Essays' in the second volume of the Green and Grose edition of 1875.

In 1758 and again in 1761 Hume went to London to see the remaining volumes of his History through the press. On the first occasion he remained there for over a year, and had serious thoughts of settling there, before deciding that he preferred the atmosphere of Edinburgh. He was well received both in high society and in literary circles, though Boswell reports Dr Johnson's saying that he once left a company as soon as Hume joined it. Johnson's 'abhorrence' of Hume did not, however, prevent them on a later occasion from being fellow guests at dinner at the Royal Chaplain's without their coming into open conflict. Characteristically, Hume used his influence with Millar to ensure the publication of the *History of Scotland* written by his friend, the Reverend William Robertson, and to promote its sales, even to the possible detriment of those of his own work. He was, however, slightly annoyed when Robertson was appointed Historiographer Royal for Scotland, in preference to himself.

At the conclusion of the Seven Years War in 1763 the Earl of Hertford, a cousin of Horace Walpole's, was appointed British Ambassador to the court of France. Being supplied with an official secretary of whose character he did not approve, he decided to employ a personal secretary and offered the position to Hume, whom he had never met. Since he was himself a very pious man, this choice was surprising, but Hume had been strongly recommended to him, as one whose name carried great prestige in France. Hume at first declined the offer, but accepted when it was renewed. He liked both Lord and Lady Hertford when he met them in London, and in October 1763 he accompanied them to Paris.

From the moment of his arrival in Paris Hume enjoyed the most extraordinary social success. As Lytton Strachey put it, 'he was flattered by princes, worshipped by fine ladies, and treated as an oracle by the *philosophes*'. His closest friends among the *philosophes* were the *Encyclopédistes* Diderot and d'Alembert, and the materialist Baron d'Holbach. There is a story of his dining at Holbach's and saying that he had never met an atheist, whereupon Holbach told him that of the persons present fifteen were atheists and the other three had not made up their minds. Among the fine ladies his chief admirer was the Countess de Boufflers, who had made herself known to him by letter in 1761. Fourteen years younger than Hume, she was the mistress of the Prince de Conti, whom she had vain hopes of marrying when her husband died. Though she never lost sight of this primary objective, she appears for a time to have been in love with Hume, and there is stronger evidence from their correspondence that he was in love with her. Though they did not meet again after Hume left Paris in January 1766, they continued for the next ten years to write to one another. His last letter to her, commiserating with her on the death of the Prince de Conti and saying of himself 'I see death approach gradually, without any Anxiety or Regret. I salute you with great affection and regard, for the last time', was written within a week of his own death.

When Hume left Paris he took Jean-Jacques Rousseau with him. Rousseau had been living in Switzerland, but his heterodox religious views had made him enemies there, nor could he rely on being undisturbed in France. Hume was persuaded, mainly by their common friend Madame de Verdelin, to take Rousseau under his protection, though warned by the *philosophes* that Rousseau was not to be trusted. Rousseau's 'gouvernante', the illiterate Thérèse Le Vasseur, was to follow, escorted by Boswell, whom she seduced on the way. At first all went well. Hume and Rousseau liked and admired one another. There was some trouble in finding a place where Rousseau would consent to live, but he finally accepted an offer from Richard Davenport, a rich country gentleman, of a house in Staffordshire at a nominal rent. Hume also arranged for him to receive an offer of a pension of £200 from King George III. But then Rousseau's paranoia broke out. Horace Walpole had written a squib against him, which Rousseau attributed to Hume. There had been jokes about him in the English press. Thérèse made mischief. Rousseau became convinced that Hume had joined with the *philosophes* in a conspiracy against him. He refused the King's pension, became suspicious of Mr Davenport, and wrote bitter letters to his friends in France, to the English newspapers, and to Hume himself. Hume tried to persuade Rousseau of his innocence, and when he failed, became anxious for his own reputation. He sent d'Alembert an account of the whole affair, giving him leave to publish it, if he thought fit. D'Alembert did publish it, together with the letters that constituted the principal evidence, and an English translation of d'Alembert's pamphlet appeared a few months later. Rousseau remained in England till the spring of 1767 and then, without a word to Mr Davenport, returned with Thérèse precipitately to France. There was no doubt that Rousseau had behaved very badly to Hume, but some of Hume's friends thought that he should have made allowances for Rousseau's paranoia and that it would have been more dignified for him not to have publicised the quarrel.

For a few months in 1765, during the interval between Lord Hertford's departure for Ireland, where he had been appointed Lord Lieutenant, and the arrival of his successor, Hume had acted as Chargé d'Affaires in Paris and shown himself to be a capable diplomat. He refused Lord Hertford's invitation to serve with him in Ireland, but in 1767 accepted an offer from the Secretary of State, Lord Hertford's brother, General Conway, to serve in London as Under-Secretary for the Northern Department. He carried out the duties of this position very successfully for the following two years.

When Hume returned to Edinburgh in 1769, he had become so 'opulent' as to enjoy an income of £1000 a year. He built himself a house in the New Town in a street off St Andrew's Square, which came in his honour to be known as St David's Street. He resumed his active social life, took no public notice of the numerous attacks that were made on his philosophy, and occupied himself with the revision of his *Dialogues Concerning Natural Religion*. The work was posthumously published, most probably by Hume's nephew, in 1779. In the spring of 1775 he was, in his own words, 'struck with a disorder in my bowels, which at first gave me no Alarm, but has since, as I apprehend it, become mortal and incurable' (D 239). He suffered little pain and never 'a moment's Abatement of my Spirits'. Boswell characteristically intruded on him to see how he was facing the prospect of death, and was convinced by his assurance that he viewed it serenely. Equally characteristically, Dr Johnson insisted that Hume must have been lying. Death finally came to him on 25 August 1776.

Hume's life largely bears out his description of himself as 'a man of mild Dispositions, of Command of Temper, of an open, social, and cheerful Humour, capable of Attachment, but little susceptible of Enmity, and of great Moderation in all my passions' (D 239). There is no doubt that Adam Smith was sincere when he concluded his obituary portrait of his friend by saying 'Upon the whole, I have always considered him, both in his life-time, and since his death, as approaching as nearly to the idea of a perfectly wise and virtuous man, as perhaps the nature of human frailty will admit.'

In histories of philosophy, David Hume is most often represented as completing a movement which was started by John Locke in 1690, with the publication of his *Essay Concerning Human Understanding*, and continued by George Berkeley, whose *Principles of Human Knowledge* appeared in 1710, the year before Hume's birth. The main theme of this movement is that men can have no knowledge of the world but what they derive from Experience, and the lines of its development are that Experience consists, as Locke put it, of Sensation and Reflection, that the operations of the mind, which are the objects of Reflection, are directed only on to the material, or their own transformation of the material, provided by the senses, and that the material provided by the senses consists of atomic elements such as colours, tactile feelings, bodily sensations, sounds, smells and tastes.

According to this story, Locke made a valiant attempt to assemble on this basis a picture of the physical world which accorded with the scientific theories of Boyle and Newton. It largely depended on his adopting a theory of perception which divided the data or, as Locke called them, 'simple ideas' of sense into two categories: ideas, such as those of solidity, figure and extension, which were not only the effects of the actions of physical objects upon our minds, but also resembled these objects in character, and ideas, such as those of colour or taste, which were nothing more than effects. These were respectively termed ideas of primary and ideas of secondary qualities. In both cases, the qualities accrued to the objects in virtue of the nature and activity of their 'minute parts,' but whereas the primary qualities actually characterised the objects, the secondary qualities were only dispositional; they were merely powers which enabled the objects, under suitable conditions, to produce ideas in us.

Berkeley, according to the view which we are considering, is held to have refuted Locke by demolishing his theory of perception. He showed not only that Locke was unjustified in drawing his vital distinction between ideas of primary and secondary qualities but, even more damagingly, that he had no warrant, on his premises, for believing in the existence of physical objects at all, that is, so long as physical objects are conceived, in the manner of Newton and Locke, as existing independently of our perception of them, and not as being simply composed of ideas or 'sensible qualities', a view which Berkeley somewhat rashly claimed to be anyhow more in accord with common sense. There had to be minds to perceive ideas, and since only a small minority of our ideas were the products of our own fancy, they had in the mass to have some external cause. But for this there was no need, no warrant, and, as Berkeley reasoned, not even any coherent possibility of having recourse to matter. God would suffice not only to cause our ideas but also to keep things in being when they happened not to enter into the field of any human perception. He might have laid a smaller burden upon God if he had taken the course, which John Stuart Mill was later to pursue, of reducing physical objects to 'permanent possibilities of sensation', and there are passages where he seems to endorse this view. But Berkeley was an Anglican bishop, and it suited his religious interests to maximise the part played by God. In Newtonian physics, provision is made for a creator to bring the universe into being, but once the machinery is started, the creator can turn his back upon it: it runs securely on its own. In Berkeley's estimation, this gave encouragement to deism or worse. He, therefore, made sure that God was constantly on the watch.

The part assigned to Hume is that of undermining Berkeley in much the same way as Berkeley had undermined Locke. Berkeley had eliminated matter, at least as the physicists conceived it, but left minds intact. Hume, an avowed sceptic, showed that this favouritism was unjustified. We had as little reason for believing in the existence of minds, as beings maintaining their identity through time, as we had for believing in the existence of material substances. There was an equal want of any rational justification for believing in the existence of Berkeley's God. But Hume carried his scepticism further. Both Locke and Berkeley had taken the concept of causality on trust. They differed only in that Locke allowed relations of force to hold between physical particles, whereas Berkeley gave minds the monopoly of causal activity. Hume set himself to analyse the relation of cause and effect, and what emerged from his analysis was that the idea of force, or of causal activity, in its ordinary interpretation, was myth. There could be no necessary connection between distinct events. All that remains, then, is a series of fleeting 'perceptions' with no external object, no enduring subject to whom they could belong, and not themselves even bound to one another.

This was the outcome attributed to Hume by the ablest of his contemporary critics, the Reverend Thomas Reid, who succeeded Adam Smith as Professor of Moral Philosophy at Glasgow University. Reid was the founder of the Scottish school of common-sense philosophers who carried the tradition into the nineteenth century, and his *Inquiry into the Human Mind on the Principles of Common Sense,* published in 1764, set the fashion of regarding the first book of Hume's *Treatise,* which Hume himself had intended his own *Enquiry* to supersede, as the primary source of Hume's philosophical views. Reid gave Hume credit for taking Locke's premises to their logical conclusion. Since the result was patently absurd it followed that something had gone wrong at the start. The principal error, as Reid saw it, was the adoption by Locke and his followers of the theory of ideas: the assumption that what is immediately perceived, whether it be called an idea, as by Locke, or a sensible quality, or, as Hume preferred, an impression, is something that has no existence apart from the perceptual situation in which it figures. If we reject this assumption, as indeed most philosophers now do, and follow common sense both in taking for granted the existence of persons to whom perceptual acts can be attributed, and in taking these persons to be directly acquainted through their senses with one and the same world of physical objects, which exist independently of being perceived, then Hume's scepticism may not have been met in every detail, but at least its most outrageous features will have been obliterated.

The same conception of Hume, as the sceptic who brought the empiricism of Locke and Berkeley to grief, appears over a century later in the work of the Oxford philosopher T. H. Green, who published an edition of the *Treatise* with a long introduction, the main purpose of which was to demolish the work that he was editing. His line of attack, however, had next to nothing in common with that of Thomas Reid. By this time, in spite of the rearguard action commanded by John Stuart Mill, the influence of Kant and Hegel was belatedly extending itself over British philosophy, increasingly to the detriment of common sense. Green was one of the leaders of this fashion, and his main objection to Hume was that he admitted no greater order into the world than what could be supplied by the mere association of ideas. Once again, Hume was held to be justified on the principles which he had inherited from Locke and Berkeley, and the moral drawn was that a new approach was needed. This had been appreciated by Kant who, in his *Prolegomena*, gave Hume credit for interrupting his 'dogmatic slumber' and giving his 'investigations in the field of speculative philosophy quite a new direction'. At Oxford in the 1930s, and perhaps still to this day in some places, the canonical view of Hume, as advanced for instance by a former Master of Balliol, A. D. Lindsay, who churned out the introductions to the handy Everyman editions of required philosophical texts, was that for all his errors and inconsistencies, which Green had relentlessly pinpointed, Hume had still performed a considerable service to philosophy. By showing on the one hand how an uncritical trust in reason had foundered in dogmatism, and on the other by reducing pure empiricism to absurdity, he had paved the way for Kant.

So far as I know, the first commentator to treat Hume neither as an appendage to Locke and Berkeley nor as a forerunner of Kant, but as a philosopher of original views which at least deserved serious consideration, was Professor Norman Kemp Smith, whose book *The Philosophy of David Hume: A Critical Study of Its Origins and Central Doctrines* was published in 1941. Kemp Smith's book is very long, and not always very lucid, but it is sustained by careful and far-reaching scholarship and it has the merit of paying close attention to what Hume actually said. For example, he points out that if Hume's principal intention had been to liquidate the estates of Locke and Berkeley, he would have been unlikely to assert, as he does in the introduction to the *Treatise*, that 'In pretending . . . to explain the principles of human nature, we in effect propose a complete system of the sciences, built on a foundation almost entirely new, and the only one upon which they can stand with any security' (T xvi). He remarks also that while Locke does figure in this introduction on the list of 'some late philosophers in England, who have begun to put the science of man on a new footing', the others whom Hume mentions, 'my Lord Shaftesbury, Dr Mandeville, Mr Hutcheson, Dr Butler' (T xvii), are all of them moral philosophers. This accords with Kemp Smith's view that Hume's main concern was to assimilate natural to moral philosophy. In moral philosophy, Hume follows Francis Hutcheson in representing our moral judgements as founded on the operations of a sovereign 'moral sense'. In natural philosophy, comprising the study of the physical world, the sovereignty passes to what Kemp Smith calls our 'natural beliefs'. These are expressions of 'feeling', itself largely governed by habit or custom and not subordinate to reason, in any strict sense of the term. It is only in the limited field of what we now call purely formal questions that reason holds sway. In sum, Hume's celebrated dictum that 'Reason is, and ought only to be the slave of the passions, and can never pretend to any other office than to serve and obey them' (T 415) was intended, on this view, to apply not merely, as has commonly been assumed, to judgements of value but to all but the purely formal exercises of our understanding. We shall have to consider later how far Kemp Smith was justified in taking this overall view of Hume's philosophy.

One point which does emerge from a careful scrutiny of the texts is the extent of the gulf between Hume and Berkeley. Admittedly, Hume speaks of Berkeley in the *Treatise* as 'a great philosopher' (T 17), but this is primarily on account of Berkeley's theory of abstract ideas, according to which 'all general ideas are nothing but particular ones, annexed to a certain term, which gives them a more extensive signification, and makes them recall upon occasion other individuals, which are similar to them' (T 17). How far this theory deserves Hume's description of it as 'one of the greatest and most valuable discoveries that has been made in late years in the republic of letters' is another question which we shall have to consider. Hume is also at one with Berkeley in rejecting Locke's distinction between ideas of primary and secondary qualities, and in the *Enquiry,* having dismissed the opinion that 'the ideas of those primary qualities are attained by *Abstraction*' as one 'which, if we examine it closely, we shall find to be unintelligible, and even absurd'

(E 154), he acknowledges his debt to Berkeley for this sceptical argument. He then goes on to say that 'most of the writings of that very ingenious author form the best lessons of scepticism, which are to be found either among the ancient or modern philosophers, Bayle not excepted' (E 155). This is a remarkable assessment of Berkeley, and made even more so by the fact, abundantly documented by Kemp Smith, that Pierre Bayle's sceptical *Dictionnaire historique et critique,* which came out in 1697, was a primary source for Hume's own scepticism. Hume was quite well aware that Berkeley would not have owned to being a sceptic. On the contrary, he bracketed sceptics with atheists and free thinkers as the adversaries whom his system was meant to frustrate. If, in spite of this, Hume maintains that all Berkeley's arguments are 'merely sceptical', his reason is 'that they admit of no answer and produce no conviction. Their only effect is to cause that momentary amazement and irresolution, which is the result of scepticism' (E 155).

Did Hume take the same view of his own arguments? The evidence, as we shall see, is conflicting, even in the famous last chapter of the first book of the *Treatise,* where Hume claims to have shown 'that the understanding, when it acts alone, and according to its most general principles, entirely subverts itself and leaves not the lowest degree of evidence in any proposition, either in philosophy or common life' (T 267–8). I shall be arguing that Hume here greatly exaggerates the sceptical import of his reasoning, but the point I now want to make is that he recoils from the conclusion 'that no refined or elaborate reasoning is ever to be received' (T 268). He does indeed remark that 'nature herself suffices' to cure him of his 'philosophical melancholy' and that he finds himself 'absolutely and necessarily determin'd to live, and talk, and act like other people in the common affairs of life' (T 269). Even so, this does not imply a rejection of philosophy. Towards the end of the chapter Hume still permits himself the 'hope to establish a system or set of opinions which if not true (for that, perhaps, is too much to be hoped for), might at least be satisfactory to the human mind, and might stand the test of the most critical examination' (T 272). And in the *Enquiry,* which we must not forget was meant to supersede the *Treatise,* the sceptical note is scarcely struck. Moreover, where it does appear, its use is positive. It serves as a weapon against superstition.

Neither is it true that Hume regarded all of Berkeley's arguments as admitting of no answer. He may have thought that there was no detectable flaw in Berkeley's disproof of the existence of matter, though even this is doubtful, but he surely believed that he had an answer to Berkeley's attempted proof of the existence of God. The answer emerges in the *Enquiry* from Hume's reply to Malebranche, who had responded to Descartes by becoming one of those who 'pretend that those objects which are commonly denominated *causes*, are in reality nothing but *occasions*; and that the true and direct principle of every effect is not any power or force in nature, but a volition of the Supreme Being, who wills that such particular objects should for ever be conjoined with each other' (E 70). Hume's comment on this excursion into fairyland, which applies equally to Berkeley, is that if we cannot penetrate the secrets of physical forces, we are 'equally ignorant of the manner or force by which a mind, even the supreme mind, operates either on itself or on body' (E 72). If we have no consciousness of this power in ourselves, we explain nothing by attributing it to a Supreme Being of which we have no idea 'but what we learn from reflection on our own faculties' (E 72). Hume also teases Berkeley by remarking that 'it argues surely more power in the Deity to delegate a certain degree of power to inferior creatures than to produce every thing by his own immediate volition' (E 71).

Hume's gift for irony matches that of his fellow historian Edward Gibbon, and like Gibbon he is most ready to display it when he writes about religion. Thus in his essay 'On the Immortality of the Soul', which we have seen that he refrained from publishing during his lifetime, he remarks that 'Nothing could set in a fuller light the infinite obligations which mankind have to Divine revelation; since we find, that no other medium could ascertain this great and important truth' (G 406). More straightforwardly, he concludes the chapter of the *Enquiry* in which he has shown that there cannot be any justification for a belief in miracles, by asserting 'that the *Christian Religion* not only was at first attended with miracles, but even to this day cannot be believed by any reasonable person without one. Mere reason is insufficient to convince us of its veracity: And whoever is moved by *Faith* to assent to it, is conscious of a continued miracle in his own person, which subverts all the principles of his under- standing and gives him a determination to believe what is most contrary to custom and experience' (E 131).

Hume is consistently hostile to Christianity, both on intellec- tual and on moral grounds. Thus in his essay 'The Natural His- tory of Religion', having conceded that 'the Roman Catholics are a very learned sect', he quotes with approval the verdict of 'Aver- roes, the famous Arabian', 'that of all religions, the most absurd and nonsensical is that, whose votaries eat, after having created, their deity' (G 343) and adds for his own part 'that there is no tenet in all paganism which would give so fair a scope to ridicule as that of the *real presence*; for it is so absurd, that it eludes the force of all argument' (G 343). Neither do the Calvinists fare any better. In their case Hume endorses the view of his friend, the Chevalier Ramsay, that their Jewish God 'is a most cruel, unjust, partial and fantastical being' (G 355). The proof of this propo- sition is given at some length and is taken to show that its adher- ents outdo the pagans in blasphemy. 'The grosser pagans contented themselves with divinizing lust, incest, and adultery; but the predestinarian doctors have divinized cruelty, wrath, fury, vengeance, and all the blackest vices' (G 356). This may be taken as a stricture only on the predestinarian doctors, but Hume had already argued in the *Enquiry* that since all human actions are determined, as much as any physical events, it follows that if they are traced back to a deity he must be 'the author of sin and moral turpitude', besides everything else. The Leibnizian idea that this is the best of all possible worlds, so that all the evil which it manifests is really a proof of God's goodness, seemed as preposterous to Hume as it did to Voltaire, through Hume lacked the temperament for writing a work of such savage irony as Voltaire's *Candide*.

In general, Hume is inclined to judge that believers in the unity of God, whether they have adopted a version of Chris- tianity, or some other type of monotheism, have made an intel- lectual advance upon the polytheists who embraced religion in what he believes to have been its most primitive forms. On the other hand, the intolerance of the monotheists, continually re- sulting in active persecution of those who dissent from their religious views, has made them 'more pernicious to society' (G 338). Hume ascribes the origin of religious belief to men's ignorance of natural causes and to 'the incessant hopes and fears, which actuate the human mind' (G 315). For the satisfaction of these hopes and the allaying of these fears they looked to beings of a similar character to themselves but endowed with much greater powers. Their invention of these beings is explained by a 'universal tendency among mankind to conceive all beings like themselves, and to transfer to every object, those qualities, with which they are familiarly acquainted, and of which they are intimately conscious' (G 317). This tendency persists even when the beings rarely or never assume corporeal form, and their number is reduced to one.

Hume nowhere proclaims himself an atheist. On the contrary, in 'The Natural History of Religion' and elsewhere in his writings, he professes to accept the Argument from Design. 'The whole frame of nature', he writes, 'bespeaks an Intelligent Author; and no rational inquirer can, after serious reflection, suspend his belief a moment with regard to the primary principles of genuine Theism and Religion' (G 309). This remark is not patently insincere and I have to allow for my own prejudices in taking it to be ironical. The fact is, however, that in the superb *Dialogues Concerning Natural Religion*, on which Hume worked intermittently during the last twenty-five years of his life, the strongest arguments, as we shall see, are put into the mouth of Philo, whose part in the dialogue is to rebut the argument from design, and I agree with Kemp Smith, to whom we own an excellent edition of the *Dialogues*, that Hume, without ever openly displaying his hand, intended the discerning reader to conclude that he adopted Philo's position. In my view, indeed, the discrediting not only of the more superstitious types of theism but of any form of religious belief was one of the principal aims of Hume's philosophy.

In reviewing the work of any of the famous seventeenth- or early eighteenth-century philosophers, one must always bear in mind that they did not draw the distinction, which has more recently arisen, between philosophy and the natural or social sciences. This is not to say that they took philosophy itself to be a special science, but rather that they regarded every form of scientific enquiry as philosophical. For them, the main division was that between natural philosophy, which concentrated on the physical world, and moral philosophy, which Hume called 'the science of human nature'. It had to be admitted that natural philosophy had advanced much the further of the two. The moral philosophers had nothing of comparable importance to set against the progress in our understanding of the physical operations of nature that began with the work of Copernicus, Kepler and Galileo, and culminated in that of Boyle and Newton. Nevertheless there was a sense in which both Locke and Hume believed that moral philosophy was dominant. Their reason, as set out by Hume in his introduction to the *Treatise*, was that 'all the sciences have a relation, greater or less, to human nature and that, however wide any of them may seem to run from it, they still return back by one passage or another' (T xv). There are sciences like those of logic, morals, criticism and politics which have a closer connection with human nature than the others, but even mathematics and the physical sciences are dependent on man's cognitive powers. 'It is impossible', Hume says, 'to tell what changes and improvements we might make in these sciences were we thoroughly acquainted with the extent and force of human understanding, and could explain the nature of the ideas we employ and of the operations we perform in our reasonings' (T xv).

Hume, as did Locke before him, sets himself to fulfil these needs. He had a qualified respect for Locke, whom he reproaches, rightly, for his excessively loose employment of the term 'idea', and also, not quite so fairly, for being betrayed by the schoolmen into mishandling the question of innate ideas, and allowing 'a like ambiguity and circumlocution' to run through 'his reasoning on ... most other subjects' (E 22); but he shared Locke's belief that the experimental method in reasoning, to which they both attributed the achievements of Newton and his predecessors, were applicable to the moral sciences. It was, however, Locke who had the deeper understanding of Newtonian theory. They both agreed that it depended upon 'experience and observation' (T xvi), but Locke noticed, as it would seem that Hume did not, that, in Newton's case, the dependence was indirect. He understood that Newton accounted for the behaviour of bodies in terms of the operations of their 'minute parts', which

were not themselves observable; and it was his attempt to re-
concile this fact with the restrictions which he laid upon the
extent of our ideas, and our consequent capacity for knowledge,
that led to much of the 'ambiguity and circumlocution' with
which Hume reproaches him. Hume, on the other hand, speaks
of Newton as though he did no more than practise straightforward induction. What Newton presumably meant by his dis-
claimer of hypotheses at the beginning of his *Principia* – the
celebrated statement 'Hypotheses non fingo' ('I do not fashion
hypotheses') – was that he advanced no propositions for which he
lacked experimental evidence. What Hume apparently took him
to have meant was that he abstained from any generalisation that
was not directly founded upon observed instances. This historical
error has some bearing upon Hume's treatment of causation, a
feature of his system to which he himself rightly attached the
greatest importance, but I shall try to show that it does not seriously diminish the force of his argument. It has no bearing upon his attempt to develop a science of the mind, since all that is at
issue here is his ability to give an accurate description of different
states of consciousness, and, remaining on the terrain of every-
day observation, to hit upon generalisations which they can
reasonably be supposed to satisfy.

On the face of it at least, the method which both Locke and
Hume pursued was very simple. They asked two questions. What
are the materials with which the mind is furnished, and what
uses can it make of them? Hume's answer to the first question is
that the material consists of perceptions, which he divides into
the two categories of impressions and ideas. In the anonymous
Abstract in which he extols the virtues of the *Treatise* he speaks
of the author as calling 'a *perception* whatever can be present to
the mind, whether we employ our senses, or are actuated with
passion, or exercise our thought and reflection' (T 647). He goes
on to say that a perception is to be called an *impression* 'when we
feel a passion or emotion of any kind, or have the images of
external objects conveyed by our senses' (T 647). '*Impressions*',
he adds, 'are our lively and our strong perceptions.' It is implied
that the reason why *ideas* are comparatively fainter and weaker is
that they come into being 'when we reflect on a passion or an
object which is not present' (T 647).

Except for the reference to external objects, which constitutes, as we shall see, a serious problem for Hume, this account of what he means by impressions is very similar to that given in the *Treatise*. He introduces them there as those perceptions which enter into our consciousness 'with most force and violence', and adds that he comprehends under the name 'all our sensations, passions and emotions, as they make their first appearance in the soul' (T 1). In the *Enquiry* he is content to say that what he means by the term *impression* is 'all our more lively perceptions, when we hear, or see, or feel, or love, or hate, or desire, or will' (E 18), but this definition is so compressed as to be misleading. The salient feature of impressions is not their force or vividness but their immediacy; this may in general have the effect of making them more lively than the images of memory or the creatures of fancy, which Hume included in the opposite camp of *ideas,* but the empirical evidence does not favour the assumption that this is always so.

Unlike Locke, Hume has no objection to saying that impressions are innate. He remarks that 'if by innate be meant, contemporary to our birth, the dispute seems to be frivolous; nor is it worth while to enquire at what time thinking begins' (E 22). In the *Enquiry*, where this passage occurs, he takes 'innate' to mean 'what is original or copied from no previous perception' and in the *Abstract*, what arises 'immediately from nature', and in both cases he concludes that all impressions are innate. This hardly seems true of our sense-impressions, according to the first definition, unless the word 'copied' is taken very narrowly. If we apply the second definition to them, we seem to have only a different way of making the point that they are the immediate contents of sense-perception. However, Hume's examples make it clear that he is concerned here chiefly with the passions. His contention is that such things as 'self-love, or the resentment of injuries, or the passion between the sexes' (E 22) are inherent in human nature.

There are passages in which Hume seems to imply that his distinction between impressions and ideas can be equated with that between feeling and thinking, but this does not mean that he is anticipating the distinction which Kant drew between intuitions and concepts. Not only do Hume's impressions enter the mind under concepts, but these concepts are guaranteed to apply to them. Hume's argument is that 'since all actions and sensations of the mind are known to us by consciousness, they must necessarily appear in every particular what they are, and be what they appear. Everything that enters the mind being in *reality* a perception, 'tis impossible that anything should to *feeling* appear different. This were to suppose, that even where we are most intimately conscious, we might be mistaken' (T 190). It is true that this statement appears in a context where Hume is concerned to deny that an impression could masquerade as an external object, but I think it sufficiently clear that he intends his assertion that 'all sensations are felt by the mind, such as they really are', to apply to our recognition of their qualities. Whether he is right on this point is an unresolved question. There is no doubt that we can honestly misdescribe our feelings, and the way things appear to us, but it can be argued that our mistakes are then purely verbal, that, as Bertrand Russell once put it, 'what I am believing is true, but my words are ill chosen'. The difficulty lies in there being cases where the line between factual and verbal error is not easily drawn. Fortunately, so far as Hume is concerned, the question can be left open. He is bound to hold that our estimates of the qualities of impressions run a relatively slight risk of error, but not that they are infallible. And indeed when it comes to our assessing the comparative proportions of spatially extended impressions, which he counts as complex, he admits the possibility of doubt and error. Our decisions as to whether they are 'greater, less, or equal' are 'sometimes infallible, but they are not always so' (T 47).

A more crucial element in Hume's account of impressions is his taking them all to be 'internal and perishing existences' (T 194). The arguments with which he supports this view are, as is frequently the case with Hume, a mixture of the logical and the experimental. The experiments which are supposed to 'convince us that our perceptions are not possessed of any independent existence' are those which typically figure in philosophical literature under the ill-chosen heading of 'the argument from illusion'. The opinion, for which there is factual evidence, 'that all our perceptions are dependent on our organs, and the disposition of our nerves and animal spirits . . . is confirm'd by the seeming encrease and diminution of objects, according to their distance; by the apparent alterations in their figure; by the changes in their colour and other qualities from our sickness and distempers; and by an infinite number of other experiments of the same kind' (T 211).

Quite apart from the question whether Hume is entitled, on his premises, to draw in this way upon physics and physiology, the argument is plainly not conclusive against opponents like Reid who take the common-sense view that in the normal exercise of sense-perception we are immediately presented with physical objects. It is not conclusive, because their position need not commit them either to denying that our perceptions are causally dependent upon a number of factors besides the existence of the object perceived, or to maintaining that we always perceive things as they really are. I think, therefore, that Hume would have done better to rely on his purely logical argument that it is self-contradictory to suppose that an entity which is defined as the content of a particular perception can lead a separate existence of its own. In short, impressions are made 'internal and fleeting' by fiat. In defiance of current fashion, I shall argue later that this, or at any rate something very like it, is a legitimate procedure, and that it can be made the basis of a tenable theory of perception.

For reasons which will soon become apparent, Hume's principal concern was with *ideas*, but the accounts which he gives of his use of the term *idea* are cursory and inadequate. In the *Treatise* he says that he means by *ideas* the 'faint images' of impressions 'in thinking and reasoning' (T 1). In the *Abstract* he says: 'When we reflect on a passion or an object which is not present, this perception is an *idea*. Impressions, therefore, are our lively and strong perceptions: *ideas* are the fainter and weaker' (T 647). Very much the same account is given in the *Enquiry*, except that the order of the explanations is reversed. The subject is introduced by the remark that 'we may divide all the perceptions of the mind into two classes or species, which are distinguished by their different degrees of force and vivacity' (E 18).

The reason why I said that these explanations are inadequate is not just that they carry the false assumption that the work of concepts, which is what ideas perform for Hume, is always carried out by images. It is rather that the emphasis is laid on the wrong factor. Let us consider, for the sake of argument, the case where the thought of some passion or sensation does take the form of an image. Now it may or may not be the case that the image is intrinsically less vivid than the passion or sensation was when it occurred. The point is that no degree of vividness or faintness can endow it with a reference beyond itself. To be an image *of* the passion or *of* the sensation, it has to be interpreted as a symbol; it has to give rise to a belief not in its own existence but in the existence of what it *represents*; and then the question of its own comparative intensity becomes irrelevant.

These points can be clearly illustrated by the example of memory. As it happens, Hume has very little to say about memory. He speaks of it as a faculty by which we repeat our impressions, though of course it is not possible, on his view, that any impression should literally be repeated, and says that what is reproduced in memory is something which, in respect of its vivacity, 'is somewhat betwixt an impression and an idea' (T 8). In general, he appears to take it for granted that memory is reliable, at least with respect to recent events. He puts the data of memory very much on a level with the immediate data of sense-perception, as sources of knowledge which serve as a basis for more venturesome inferences, and we shall see that it is with our right to make these more venturesome inferences that he is primarily concerned. What is of interest in the present context is the way in which Hume distinguishes between ideas of memory and ideas of the imagination. According to a fundamental principle of Hume's, which we shall presently examine, both must be derived from previous impressions, but whereas memory, so long as it is functioning properly, 'preserves the original form in which its objects were presented', as well as the order in which they originally occurred, the imagination, so long as it remains within the realm of past impressions, is free to arrange their counterparts in any order and combine them in any way it pleases, whether or not these combinations have actually occurred. There is, however, the difficulty, for the most part ignored by Hume, but mentioned once in passing, that we cannot return to our past impressions to discover where this difference obtains. Consequently, the only way we actually have of distinguishing memory from imagination 'lies in its superior force and vivacity' (T 85); the same terms as are used in the *Enquiry* to distinguish impressions from ideas. By comparison with the 'lively and strong' ideas of memory, the ideas of the imagination are 'faint and languid' (T 9).

This is clearly unacceptable. In the first place, it is perfectly possible to remember a past experience, say a conversation in which one took part, or one's feeling of disappointment when something went wrong, without the aid of any images at all. It is possible even to imagine that something is the case, and yet dispense with any actual image. Let us, however, consider only the cases where images do play a role. It is simply false that images which serve the imagination are always fainter and more languid than those that enter into memories. Indeed, Hume himself admits this in one of the sections of his Appendix to the *Treatise*, and has to take refuge in saying that even if a poetical fiction paints the livelier picture, still the ideas it presents 'are different to the *feeling* from those which arise from the memory and judgement' (T 631). But this is really to concede that the whole question of relative vividness is beside the point. The distinctive feature of memory, whether it is served by images or not, is that it exhibits one's belief in something which one has learned: in the sense with which Hume is concerned, the belief is a belief in the occurrence of some past experience. The distinctive feature of the imagination, except in the sense in which to say that something is imaginary is to imply that it does not exist, is that it is neutral with regard to the existence of the states of affairs which it represents. Hume was right in so far as there is only a difference of 'tone' between one's memory of a past experience and one's imaginative reconstruction of an experience which one truly believes to have occurred, but he was mistaken if he took this to be the only or even the main point at issue.

I use the conditional because it is not clear that he always did so. Hume was not a consistent writer and in the Appendix to the *Treatise*, in a note on the section entitled 'Of the Nature of the Idea, or Belief', he says that in giving a philosophical account of belief 'we can do no more than assert, that it is something *felt* by the mind, which distinguishes the ideas of the judgement from the fictions of the imagination' (T 629). It is true that he does not explicitly include memories among the ideas of the judgement, but this can be put down to the context; and it is also true that he goes on to speak of the ideas to which we assent as being 'more strong, firm and vivid' than our reveries. But now it looks as if words like 'vivid' are being used in a technical sense which implies that the ideas which they qualify command assent.

The trouble was that Hume lacked an adequate theory of meaning and reference. Towards the end of his Appendix to the *Treatise* he admits to being dissatisfied with his treatment of belief. Since he associates belief primarily with inference, and assumes that all inferences must ultimately be based on some impression, he begins by defining belief as 'a lively idea related to or associated with a present impression' (T 96), but here again the word 'lively' is out of place unless it is simply cover for some such expression as 'assented to'. Hume did indeed make the valid point that belief cannot be a further idea, tacked on to the idea which gives the belief its content; for, as we should now put it, what a sentence is used to state remains the same, whether it is believed or not. Neither, Hume argues, can it be an impression, because it attaches to conclusions which consist only of ideas. He sometimes calls it a sentiment, without making it clear what he takes a sentiment to be if it is neither an impression nor an idea. In the end, he is reduced to saying in the *Enquiry* 'that belief consists not in the peculiar nature or order of ideas, but in the manner of their conception, and in their feeling to the mind' (E 49). This is scarcely illuminating, but in fairness to Hume it can be said that the problem of giving an analysis of belief which is neither trivial nor circular still awaits solution. Recent attempts to analyse it in terms of propensities to action have not, to my mind, been successful.

It might be thought that a theory of reference was contained in Hume's treatment of abstract ideas, but here, more than anywhere else, he is handicapped by his false assumption that the use of a concept consists in the framing of an image. This leads him to expend unnecessary labour in proving that images have determinate qualities, a proposition which he does not clearly distinguish from the proposition, for which he also argues, that images represent particular individuals which themselves have determinate qualities. I am not sure that the first of these propositions is true, but even if they are both true they are irrelevant, since the use of a general term need not be accompanied either by an image or by the thought of any particular individual. Hume is, indeed, right in welcoming Berkeley's 'discovery' that terms are made general not by their standing for an abstract entity, but by their use, but he has nothing more informative to say about their use than that 'individuals are collected together and placed under a general term with a view to that resemblance which they bear to each other' (T 23).

I referred earlier to a fundamental principle of Hume's concerning the derivation of ideas. This principle, as originally stated in the *Treatise*, is '*That all our simple ideas in their first appearance are derived from simple impressions, which are correspondent to them, and which they exactly represent*' (T 4). The extreme importance of this principle for Hume is that it provides him with a criterion of the legitimacy of ideas or concepts. As he puts it in the *Enquiry*, 'When we entertain any suspicion that a philosophical term is employed without any meaning or idea (as is but too frequent) we need but enquire, *from what impression is that supposed to be derived?*' (E 22).

Hume treats this as an empirical generalisation, to which, strangely enough, he at once suggests a counter-example. He envisages a case in which a man has been acquainted with various shades of colour, but has missed out one of them, and suggests, quite correctly, that the man could form an idea of the missing shade as being intermediate in hue between two of the others. Having devised this counter-example, Hume wantonly dismisses it, saying that it is too 'particular and singular' to oblige him to give up his general maxim; and indeed he continues to treat his maxim as though it held universally. He could have avoided this affront to logic if he had modified his principle so as to make it apply not to the origin but to the realisation of ideas. It would then require of an idea that it be capable of being satisfied by some impression. This principle can, I think, be defended if the word 'satisfied' is construed generously, so as to allow an idea to be satisfied if it enters into a theory which can be confirmed, at least indirectly, in sense-experience. This would, however, take us rather far from Hume. A modification which he might have accepted is that ideas must be capable of being instantiated by impressions, but this, though it accommodates his counter-example, is still too restrictive. In practice, Hume accepts such ideas as can be instantiated by bodies, which are themselves developed out of impressions by certain activities of the imagination. We shall be discussing the question how far this practice is at variance with his official theory.

Having fixed his attention on ideas, Hume proceeds to review the ways in which they are associated. In the *Treatise* he distinguishes seven different kinds of philosophical relation, which he divides into two groups according as they can or cannot be 'the objects of knowledge and certainty'. The four that can attain this mark, because they depend solely upon the intrinsic properties of ideas, are *resemblance, contrariety, degrees in quality* and *proportions in quantity or number*. The remaining three are *identity, relations of time and place*, and *causation*. Hume's reason for placing *identity* in the second group is that objects which are perfectly resemblant may still be numerically different if they do not coincide both in space and time. This confirms, what was already clear, that Hume's so-called relations between ideas are not merely such, but also extend to relations between the objects which fall under the ideas. Indeed, of all the relations which he lists, perhaps only that of *contrariety* is purely conceptual.

This division between relations prefigures the second of Hume's fundamental principles, which is that 'All the objects of human reason or enquiry may naturally be divided into two kinds, to wit, *Relations of Ideas*, and *Matters of Fact*' (E 25). In this proposition, which is taken from the *Enquiry*, affirmations concerning relations of ideas are treated as being purely conceptual, and consequently as being 'either intuitively or demonstratively certain'. They include the sciences of geometry, algebra and arithmetic. Hume has next to nothing to say about them, his own interest lying in our beliefs about matters of fact. This leads him, in the *Enquiry*, to reduce his list of relations to three, 'namely *Resemblance, Contiguity* in time and place, and *Cause and Effect*' (E 24). The first two are merely principles of association. The important relation is that of *Cause and Effect*, on which, as Hume puts it, all reasonings concerning matters of fact seem to be founded.

It is this relation, then, that Hume sets out to explore. It must, however, be noted that when he does explore it, in the domain of natural philosophy, he takes it as holding not between fleeting impressions but between enduring objects. This is shown by his examples, and indeed it is only if it is taken in this way that his analysis of causation can appear at all plausible. Equally, when it comes to moral philosophy, he relies on the existence of persistent selves. We must, therefore, begin by showing how he can arrive at these conceptions without too flagrant a display of inconsistency.

Whatever scepticism Hume may have professed, there is no doubt that he believed in the existence of what may be called the physical objects of common sense. Even if we assume that he left his philosophical self behind, as he in fact did not, when he wrote his volumes of history and his political essays, and, as he says he did, when he played backgammon and was merry with his friends, there is abundant evidence of such a belief in his philosophical writings themselves. He shows interest in the passions and emotions, which he enlists under the heading of 'secondary impressions', because they have a part to play in his moral theory, but neither in the first book of the *Treatise*, nor in the *Enquiry concerning Human Understanding,* does he pay very much attention to the original impressions of sense. He makes a number of remarks about them in general, referring, as we have seen, to the force and liveliness with which they are supposed to strike upon the mind, or to their being 'internal and perishing existences', but he has little to say about them in detail. When he adduces examples of matters of fact, he refers not to colours and shapes, but to the death of Caesar in the Senate-house at Rome, to the properties of mercury and gold, to the muscles and nerves of human bodies, to the sun and the planets, to flowers and trees, to the impact of billiard balls upon one another. It is on the constant conjunction of objects of this sort that, as we shall see, his analysis of causation depends. It would have been no use his applying his theory to impressions, as he must have done if he had really believed that there was nothing else available, for the very good reason that our actual impressions do not display the requisite regularity. To take a simple example, it may well be the case that roses planted in the spring habitually flower in the summer, but this is not to say that an impression of seeing a rose planted is habitually followed by an impression of seeing it come into flower. One may not be there at the later date, and even if one is there, one's attention may be otherwise occupied.

Let it be allowed, then, that Hume believed, and for his own philosophical purposes needed to believe, in the existence of bodies. In fact, at the beginning of the section of the *Treatise* entitled 'Of Scepticism with regard to the Senses', which supplies the main source for attributing to him a 'theory of the external world', he himself makes the cryptic statement 'We may well ask, *What causes induce us to believe in the existence of body?* but 'tis vain to ask. *Whether there be body or not?* That is a point, which we must take for granted in all our reasonings' (T 187). Even so a number of questions remain to be answered. We have seen that the bodies the existence of which Hume took for granted were such things as houses and books and trees and the bodies of animals and human beings, including their inner organs. How did he conceive of them? Like Locke, as the external causes of sensory impressions? Like Berkeley, as congeries of sensible qualities? Or after some fashion of his own? He has an answer for the causal question how we come to believe in the existence of body, but what does he mean by saying that it is vain to ask whether there be body or not? The most obvious interpretation to give to this statement would be that our belief in the existence of body is unquestionably true. Yet the upshot of Hume's investigation of the causes of this belief is that, whether it be considered in its vulgar or in what he calls its 'philosophical' form, it is not only not true but thoroughly confused. Admittedly, he does not expect this conclusion to carry any lasting conviction. 'Carelessness and in-attention', he said, 'alone can afford us any remedy. For this reason I rely entirely upon them: and take it for granted, whatever may be the reader's opinion at this present moment, that an hour hence he will be persuaded there is both an external and internal world' (T 218). Should we take this as confirming Kemp Smith's view that Hume intended our 'natural beliefs' to triumph over our understanding? But in what exactly were these natural beliefs supposed to consist?

Once more, the evidence is conflicting. There are passages which seem to imply that when Hume speaks of 'external' objects, he conceives of them, in Locke's way, as relatively enduring causes of our 'internal and fleeting' sense-impressions. For instance, we have already seen that, in the course of defining impressions in the *Abstract,* he speaks of our having images of external objects conveyed by our senses. Again, in the *Treatise,* at the beginning of the Book 'Of The Passions', where he makes his distinction between original and secondary impressions, he speaks of original impressions as 'Such as without any antecedent perception arise in the Soul, from the constitution of the body, from the animal spirits, or from the application of objects to the external organs' (T 276). In the chapter which he devotes to our ideas of space and time, in the first book of the *Treatise*, he says that 'we can never pretend to know body otherwise than by those external properties, which discover themselves to the senses' (T 64), and the implication that bodies have other properties, which lie beyond our knowledge, is confirmed by a note on the chapter, in one of the *Appendices*, where he speaks of the real nature of the position of bodies, which might or might not allow for a vacuum, as being unknown (T 639). He takes this to be implied by the 'Newtonian philosophy' which for the most part he does not contest. On the other hand, in the one place in which he examines its foundations, he does very firmly contest it; and he also implies that it is not what the carelessness and inattention of his reader will allow him to accept.

As is so often the case with Hume, the *Enquiry* summarises with greater elegance and clarity a position for which the arguments are developed in the *Treatise*. In this instance, we are told that

It seems evident, that men are carried, by a natural instinct or pre-possession, to repose faith in their senses; and that, without any reasoning, or even almost before the use of reason, we always suppose an external universe, which depends not on our perception, but would exist, though we and every sensible creature were absent or annihilated. (E 151)

Hume goes on to ascribe this belief even to 'the animal creation'. He then adds:

It seems also evident, that, when men follow this blind and powerful instinct of nature, they always suppose the very images, presented by the senses, to be the external objects, and never entertain any suspicion, that the one are nothing but representations of the other. This very table, which we see white, and which we feel hard, is believed to exist, independent of our perception, and to be something external to our mind which perceives it. Our presence bestows not being on it:

our absence does not annihilate it. It preserves its existence uniform and entire, independent of the situation of intelligent beings, who perceive or contemplate it. (E 151–2)

Unfortunately, in Hume's view, 'this universal and primary opinion of all men' does not withstand critical examination. It succumbs to 'the slightest philosophy' with the result that 'no man, who reflects, ever doubted that the existences, which we consider, when we say, *this house* and *that tree*, are nothing but perceptions in the mind, and fleeting copies or representations of other existences, which remain uniform and independent' (E 152). This would be all very well if we had any good reason to believe in the existence of these independent objects, but Hume maintains that we have not. The supposition that our perceptions are connected with external objects can evidently not be founded in experience and is also, Hume says, 'without any foundation in reasoning' (E 153). This leaves the sceptic in command of the field. Hume is not happy with this result, but does not look for any way of contesting it, such as searching for some flaw in the argument. What he does instead is simply to shrug it aside. We set the extreme sceptic at defiance by asking what purpose he thinks that his activity serves. He 'cannot expect, that his philosophy will have any constant influence on the mind: or if it had, that its influence would be beneficial to society' (E 160). On the contrary, it would be very harmful, since the inaction to which it would lead would put an end to men's existence. But 'Nature is too strong for principle.' The sceptic will be brought to confess that 'all his objections are mere amusement, and can have no other tendency than to show the whimsical condition of mankind, who must act and reason and believe' (E 160). Remarks of this sort do, indeed, lend support to Kemp Smith's theory that Hume was concerned to show that reason ought to give way to our natural beliefs, but I do not see them as establishing it decisively. If this had been Hume's intention, I do not think that he would have concluded the section on a sceptical note, stressing our inability to justify the assumptions on which we act or to remove the objections which may be raised against them. Neither does he explicitly assert that our natural beliefs are true.

There is, however, an important point on which I agree with Kemp Smith; and that is that it is possible to elicit from Hume's line of reasoning a theory of perception which does not dissolve into scepticism. It will mean parting company with Hume at one or two places, but the divergencies will not be so great as to prevent our arriving at a fair accommodation. They will consist for the most part in a re-evaluation rather than a rejection of his arguments. I shall try to show presently how I think that this can be done, but first I want to retrace the course that he actually pursues in the relevant sections of the *Treatise*, of which the most important for our present purpose is that 'Of Scepticism with regard to the Senses'.

The main questions which Hume sets out to answer in this section are 'Why we attribute a *continu'd* existence to objects even when they are not present to the senses; and why we suppose them to have an existence *distinct* from the mind and perception?' (T 188). He treats these questions as interdependent, in the sense that to answer either one of them would be to answer the other. He is right to the extent that if objects did have a continued existence, as he defines it, they would also have a distinct existence, but the converse need not hold. It might be and indeed has been maintained, for example by Bertrand Russell in *The Problems of Philosophy* and elsewhere, that the objects which are immediately present to the senses are distinct from the mind and yet have only a momentary existence, because of their causal dependence on the bodily state of the percipient.

On this point I side with Hume, with one important reservation. As we have noted earlier, Hume, while mainly relying on empirical arguments, including the reference to causal factors, shows some inclination to make it a truth of logic that sense-impressions are inseparable from the perceptual situations in which they occur. I remarked that the empirical arguments are not conclusive, though, as we shall see, they do present an obstacle to the currently bland acceptance of a position like Reid's, and I suggested that Hume should have been content to stand on his logical principle. My reservation is that this does not commit him to defining impressions from the outset as being dependent on the mind. In fact such a definition would be inconsistent with his own view, which we have yet to examine, that the self is nothing but a bundle of logically independent perceptions, from which he infers that a perception could exist apart from anything else. This is not, however, the point that I wish to stress, since he does not maintain that any perceptions actually do so. My point is rather that if impressions are to be taken as primitive in the order of knowledge, they cannot be accorded an initial dependence on either mind or body. At this stage neither minds nor bodies enter into the picture.

But is Hume entitled to take impressions as primitive? I maintain that he is, for the two following reasons. First, it is evident that one's ground for accepting any proposition concerning what Hume calls a matter of fact must finally depend on the truth of some judgement of perception. Secondly, it is easy to show that our ordinary judgements of perception assert more than is vouchsafed by the sense-experiences from which they issue. Though one is not normally conscious of making any inferences when one ventures on such a simple perceptual judgement as 'This is an ashtray' or 'That is a pencil', there is a sense in which they do embody inferences. But then these inferences must have some foundation. There must, in Bertrand Russell's terminology, be 'hard data' on which they are based. And Hume's impressions are simply these hard data, called by another name.

The second step in this argument has been contested, but to see that it is valid one needs only to be reminded of the very wide range of assumptions which our ordinary judgements of perception carry. For a start, there are the assumptions involved in characterising anything as a physical object of an observable kind. It has to be accessible to more than one sense and to any suitably equipped observer. It has to be capable of existing unperceived. It has to occupy a position or series of positions in three-dimensional space and to persist throughout some period of time. Neither are we normally committed only to these general assumptions. It is seldom that we are content to make so weak a claim as that we are perceiving a physical object of some sort or other. In the normal way, we identify it more specifically as an ashtray or a pencil or a table or whatever, and thereby commit ourselves to a set of further assumptions. These may relate to the object's origins, as when we claim to perceive some living thing, or to its physical constitution. In referring to an artefact, like a pencil, we make an assumption about its causal powers. Very often our description of an object which we claim to see or touch carries implications about its possible effects on the other senses, about the sound or taste or smell which we accord it the power to produce.

But now it is surely obvious that this whole wealth of theory cannot be extracted from any single occasion of sense-experience. I see what I do not hesitate to identify as a reading lamp on the table in front of me, but there is nothing in the visual pattern, considered on its own, from which to deduce that the object is tangible, that if any other observer were present he also would see it, that it will remain unperceived, most probably in the same position, that it is partly made of brass, that there is a socket for an electric bulb under its shade, and that this bulb can be made a source of light. What is present to my senses, so far as the lamp is concerned, is just a visual pattern, and all the rest is inference. This does not mean that I speak falsely when I say that I see the lamp or even that I am misusing the verb 'to see'. No doubt the assumptions which I am tacitly making are true in this instance, and it is common practice in the employment of perceptual verbs to take as their accusatives, not what I have called hard data, that is to say the actual contents of the sensory experiences in question, but the objects to the presence of which the hard data serve as sensory clues. But to overlook the hard data is not to abolish them, and the inferences are not removed by being unacknowledged. The common run of contemporary philosophers who are content to take the perception of physical objects as their starting-point need not be in error, since no one is obliged to embark on an analysis of perception. They fall into error only when they maintain, or imply, that no such analysis is possible.

The standard objection to the adoption of Hume's starting-point is that it imprisons the subject in a private world, from which he cannot ever escape. This objection would be serious if it were valid, but it is not valid. There is nothing private about the description of sensory patterns. It can be understood by anyone who has the requisite experiences. It is true that the patterns are made concrete by their occurrence in a particular sense-field at a particular time, and that this sense-field will in fact figure in the experience of only one person. It is equally true that a person's act of perceiving some physical object at a particular time is his and not anybody else's. The crucial point is that no reference to their ownership enters into the definition of impressions. As primitive elements, they are neutral in this and every other respect, except that of their intrinsic character. It remains to be seen whether we can find a passable route from them to physical objects, and whether we can then find a means of making a distinction between these physical objects and our perceptions of them. It may turn out in one case or the other that we shall come upon insurmountable obstacles, but this is certainly not a question that can be prejudged at the start.

For Hume the obstacles come very early. We have already seen that he takes it to be certain 'that almost all mankind, and even philosophers themselves, for the greatest part of their lives, take their perceptions to be their only objects, and suppose that the very being, which is intimately present to the mind, is the real body or material existence' (T 206). They must, however, be mistaken, since it is characteristic of what he here calls real bodies that they have a continued and distinct existence, whereas perceptions are 'dependent and fleeting'. Thus they are guilty of the contradiction of supposing that one and the same object both does and does not persist through time.

Before examining Hume's account of the way in which we are seduced into this error, it is worth remarking that, on his own showing, the contradiction is not so flagrant as it may at first sight appear. We have seen that he includes identity in his list of philosophical relations, but it now turns out to be a relation that no perception satisfies. This conclusion emerges from Hume's treatment of the question how we can come by the idea of identity. It cannot, he argues, be conveyed by one object, since the proposition 'an object is the same with itself' would be meaningless 'if the idea expressed by the word, *object*, were in no way distinguish'd from that meant by *itself*' (T 200). The idea which a single object conveys is not that of identity but that of unity. But if a single object cannot convey the idea of identity, it is even more obvious that a multiplicity of different objects cannot. Where then can the idea of identity come from? In what can it possibly consist?

Hume's answer to these questions is that the idea of identity is the product of a mistake into which we naturally fall when we think about time. He has previously argued that 'time, in a strict sense, implies succession, and that when we apply its idea to any unchangeable object, 'tis only by a fiction of the imagination, by which the unchangeable object is suppos'd to participate of the changes of the co-existent objects, and in particular of that of our perceptions' (T 200–1). We are, however, firmly addicted to this play of the imagination, which enables us to keep the idea of an object poised, as it were, between those of unity and number. So when we attribute identity to an object, what we must mean is 'that the object existent at one time is the same with itself existent at another' (T 201). Provided that they occurred in constantly changing circumstances, this could be true of at least some of our perceptions, if they themselves persisted unchanged. The trouble is that they do not.

How then do we come to think that they do? Hume's explanation, in his own words, is

that all those objects, to which we attribute a continu'd existence have a peculiar *constancy*, which distinguishes them from the impressions, whose existence depends upon our perception. Those mountains, and houses, and trees, which lie at present under my eye, have always appear'd to me in the same order: and when I lose sight of them by shutting my eyes or turning my head, I soon after find them return upon me without the least alteration. My bed and table, my books and papers present themselves in the same uniform manner and change not upon account of any interruption in my seeing or perceiving them. This is the case with all the impressions, whose objects are suppos'd to have an external existence; and is the case with no other impressions, whether gentle or violent, voluntary or involuntary. (T 194–5)

What this comes to is that the close resemblance between successive impressions and, what is equally important, their standing in apparently constant spatial relations to the members of series which exhibit the same internal resemblances, leads us to identify them with one another, and to ignore the interruptions which in fact occur between them. The result is that they are replaced in our imagination by a persistent thing, indifferently referred to by Hume as an object or a perception, which we conceive of as existing when it is not being perceived. Since Hume has shown some inclination, as we have seen, to make it a logical truth that impressions are internal and fleeting, it is something of a surprise to find him saying that 'the supposition of the continu'd exist-

ence of sensible objects or perceptions involves no contradiction'
(T 208). His reason is not only that 'every perception is distinguishable from another, and may be considered as separately existent', with the consequence that 'there is no absurdity
in separating any particular perception from the mind' (T 209).
He also concedes that 'the same continu'd and uninterrupted
Being may be sometimes present to the mind, and sometimes
absent from it, without any real or essential change in the Being
itself' (T 207). I think that this concession is genuine, and not
just a description of the reach of our imagination, though I shall
presently try to show that he did not need to make it in this
particular form. There is a way in which he could have obtained
what it gives him, while still making the denial of a continued
existence to any actual impression a matter of logic rather than
empirical fact. As it is, he holds it to be simply false that any
perception continues unperceived, and consequently false that
interrupted perceptions can be identical. His ground for this
conclusion lies in his acceptance of the dubious argument from
illusion. He reasons correctly that if perceptions had a continued
existence they would also have a distinct existence, but then
argues that a distinct existence is denied to them by experience.
As given to sense, motion and solidity, colours, sounds, heat and
cold, pain and pleasure are all on the same footing. They are
'nothing but perceptions arising from the particular configura-
tions and motions of the parts of body' (T 192–3).

Hume accords the phenomenon of constancy the primary role
in causing the imagination to transform impressions into enduring objects, but he does not think it sufficient on its own. It is
aided in its work by what he calls coherence, of which he gives
two examples. The first example is that of his returning to his
room after an hour's absence to find that his fire is burning less
brightly. Since he has frequently witnessed the process of his
fire's dying down, his imagination fills the gap. The second
example is more complex. It is that of a porter bringing him a
letter from a friend two hundred leagues away. He hears the
noise of the door turning on its hinges but does not see it. He
does not see the porter climbing the stairs. Neither on this occasion has he observed the action of the post and ferries which
brought the letter to him. Nevertheless he has come, through past

experience, to associate the creaking of the door with the sight of its opening; he has learned that porters do not appear in his room without having climbed the stairs to get there; experience has taught him that letters do not arrive from distant places without being conveyed by observable means of transport. So once again his imagination fills in the gaps. It is, however, to be noted that his imagination is here given far more work to do than in the previous example. It is no longer just a matter of its supplying the missing counterparts to the members of previous resemblant series. There may very well not have been any such series; for example, it is not very likely that Hume had ever actually witnessed the continuous movement of a letter over a distance of two hundred leagues. In this instance his imagination has not only extended the principle of constancy so as to supply itself with a sufficient stock of ordinary objects. It has also imposed, as Hume himself admits, a greater degree of coherence in their operation than has actually occurred in his past experience. What makes this procedure legitimate is its explanatory force.

Even so, Hume maintains that our imagination is deceiving us. The fact is that our percepts do not have a continued and distinct existence and that is that. Philosophers try to get over the difficulty by drawing a distinction between perceptions and objects, allowing the perceptions to be 'interrupted and perishing', and attributing to the objects 'a continued existence and identity'. But this 'new system' is a fraud. 'It contains all the difficulties of the vulgar system, with some others, that are peculiar to itself' (T 211). In the first place, it owes its appeal to the imagination entirely to the vulgar system. 'Were we fully convinced that our resembling perceptions are continu'd, and identical, and independent, we shou'd never run into this opinion of a double existence' (T 215), for it would then serve no purpose. Neither should we adopt it if we were fully convinced that our perceptions were dependent and interrupted, for our search for something to which to ascribe a continued existence would then lack any motive. It is the vacillation of our minds between two contrary hypotheses that leads philosophers to accept them both while trying to disguise the contradiction. Secondly, this hypothesis finds no support in reason, for we can have no rational ground for correlating our perceptions with objects which *ex hypothesi* are never themselves experienced. What the philosophers, who engage in this fraud, are really doing is to invent a second set of perceptions to which they attribute the continued and distinct existence which they would be attributing to our actual perceptions if their reason did not forbid it. In short, the philosophical system is 'loaded with this absurdity, that it at once denies and establishes the vulgar supposition' (T 218).

I am going to argue that Hume does less than justice to the philosophical system. It is untenable as it stands, but it does point the way to a solution of his difficulties. He is, however, justified in his onslaught on the Lockean version of it, which was presumably his actual target. Whatever distinction we end by drawing between things as they really are, and things as they appear to us, it cannot legitimately take the form of a multiplication of 'worlds'. Physical objects cannot enter the scene merely as the imperceptible causes of what one perceives, and if they could so enter it there would be no warrant for endowing them with any resemblance to their perceptible effects.

How then do we come by a warranted belief in the existence of physical objects having the perceptible properties with which we normally credit them? I suggest that Hume himself supplies the best answer. The phenomena of constancy and coherence, which he invokes to explain how we fall into the illusion of supposing our perceptions to have a continued and distinct existence, can be regarded instead as affording an adequate basis for an imaginative transformation of sense-impressions, or percepts, as, following Russell, I prefer to call them, into the constituents of the physical world of common sense. The only serious error that Hume made was to assume that the resulting objects were fictions. What there is depends in part on what our theories allow there to be, and the theory which can be developed out of percepts, on Hume's principles, contains acceptable criteria of existence. There is no reason, therefore, for denying that the objects which satisfy these criteria do in fact exist.

So far as I know, the first philosopher to make a serious attempt to legitimise the vulgar belief in physical objects, by the development of Hume's notions of constancy and coherence, was H. H. Price, whose book *Hume's Theory of the External World* was published in 1940. It is a characteristically thorough and ingenious work, which has not received the attention that it merits. Price borrows from Russell the term 'sensibilia' to refer to impressions which may not be actually sensed, and shows in convincing detail how the occurrence of gaps at different places in the various series of impressions that we take as emanating from the same physical object can sustain the conception of unsensed sensibilia as filling in these gaps. He does not think that Hume would have attached any meaning to the statement that these unsensed sensibilia really existed, but he does think that Hume could consistently have adopted either the theory that when we refer to some physical object we are asserting that our actual impressions are such as they would be if the appropriate sensibilia existed, or the more pragmatic theory that statements about physical objects have no truth-value, but are to be construed as more or less successful recipes for explaining and predicting the occurrence of actual impressions. I think that a more realistic theory would be preferable to either of these alternatives, though it may be that Price's suggestions preserve a closer fidelity to Hume. My main objection to Price's argument it that it represents physical objects as what he calls 'families' of sensibilia, in a sense which allows for a family, the various members of which have conflicting properties, to occupy, or be as if it occupied, a particular place at a particular time. He tries to avoid the contradiction by relativising properties to points of view, so that a sensibile which is elliptical from one point of view may be spatiotemporally coincident with a sensibile, or an actual impression, which is round from a different point of view, but this is a desperate device and it may not even be coherent.

I think that we can arrive at a simpler as well as a more realistic theory if we take advantage of the fact that visual and tactual impressions occur in sense-fields which are spatially extended and sensibly overlap in time. In other words, our hard data include not only individualised patterns but the spatial and temporal relations which they bear to one another. This is a feature of our experience to which Hume pays surprisingly little attention. He does, indeed, devote five of the six sections which compose Part II of the first Book of the *Treatise* to the topic of space and time, but they are mainly concerned with conceptual difficulties which he might more profitably have left to the mathematicians, had he been already disposed to consign geometry to them, and to maintain the sharp distinction which he was to draw in the *Enquiry* between questions concerning relations of ideas and those concerning matters of fact. As it is, he insists on too simple a derivation of mathematical concepts from sensory impressions and is consequently driven to repudiate the notion of infinite divisibility. He argues, instead, that the idea and therefore the impression of any finite extension must be compound, consisting of the juxtaposition of a finite number of mathematical points. These points are concrete objects, being either coloured or tangible, according as they are susceptible to sight or touch. They are *minima sensibilia*, not in virtue of being the smallest objects that we are capable of sensing, since the invention of more powerful microscopes might show that an impression which we had thought to be simple was really compound, but in virtue of the fact that they have no parts. From this Hume infers that every visual or tactual sense-field is a plenum, since we can have no idea of a spatial interval between two impressions except as something compounded out of these coloured, or tangible, indivisible points. What is true of space applies equally to time, with the same terms in each case and the substitution of immediate succession for temporary adjacency. Among other obvious difficulties, Hume's theory clearly falls foul of Zeno's paradoxes, but the theory of mathematical continuity was not properly developed until the nineteenth century, and Hume's troubles partly arose from his inability to see how an infinite number of parts could constitute anything less than an infinite whole. Even so it is strange that the only spatio-temporal relation which he was willing to acknowledge as given to sense was that of contiguity, for this is to impose a restriction upon our sense-experience to which it simply does not conform.

The spatio-temporal continuity which actually characterises most of our sense-experience, and in particular the sequence of our visual sense-fields, sustains the projection of spatial and temporal relations beyond the limits in which they are originally given. In this way successive sense-fields can come to be regarded as spatially adjacent. Then the facts which Hume summarises under his headings of constancy and coherence, the appearance of similar impressions in similar sensory surroundings, make it natural, as he remarked, for the observer to adopt a new measure of identity, according to which these corresponding impressions are not merely similar but identical. Our only difference from Hume at this point is that he represents the observer not as adopting a new measure of identity but as making an error of fact. In the same way, Hume's account of the process by which the fact that impressions are systematically 'recoverable' leads to their being thought of as persisting unsensed, should be taken, I suggest, not as the explanation of another error, but as a ground for extending the concept of identity. It is, on this view, an entirely legitimate exercise of the imagination that visual impressions, which appear successively, should be conceived to exist simultaneously and to occupy permanent positions in an indefinitely extended three-dimensional visual space.

At this point, however, we need to go a little beyond Hume. The actual impressions, which we assign to the same object, are not, as he wrongly says, exactly resemblant. Even if we make no allowance, as yet, for changes in the object, there will be variations resulting from changes in the state or position of the observer. What we must, therefore, imagine as persisting is not any actual impression but what I have called elsewhere a 'standardised percept'.* This is a synthesis of what Price calls nuclear sensibilia, that is, such as are obtainable from optimal points of view. It serves as a model which actual impressions match more or less closely. These standardised percepts may also be described as visual continuants. They can be regarded as undergoing qualitative changes when, under a variety of conditions, the pattern appears to vary in one or more aspects, while the remainder stay constant in a predominantly constant sensory environment.

I can here only summarise the successive stages through which the theory may now be supposed to pass. First, places are conceived in detachment from their occupants, which allows for the possibility of movement. Then, for a variety of reasons, a set of visual continuants is picked out as forming what we may call the 'central body', a term borrowed from the great American pragmatist C. S. Peirce. This is, of course, the observer's own body, though not yet characterised as such. Tactual space is constructed on much the same principle as visual space and grounds

* In *The Central Questions of Philosophy*.

are discovered for attributing tactual qualities to visual continuants. Sounds and smells and tastes are brought into the picture by their being traced to their apparent sources. The observer begins to make some simple causal correlations, which provide him with an outline of the way things are. Most of his impressions fit into this scheme but some do not. Self-consciousness arises with the identification of other visuo-tactual continuants as resembling the central body in being the sources of signs. Most of these signs can be interpreted as corroborating the main body of the observer's experiences, but again some cannot. This is the source of the distinction between what is public and what is private. In the final stage, if I may quote my own account of it, the visuo-tactual continuants 'are cut loose from their moorings. The possibility of their existing at times when they are not perceived is extended to the point where it is unnecessary to their existence that they ever should be perceived, or even that there should be any observers to perceive them. Since the theory also requires that these objects do not change their perceptible qualities except as a result of some physical alteration in themselves, they come to be contrasted with the fluctuating impressions that different observers have of them. In this way the objects are severed from the actual percepts from which they have been abstracted and are even regarded as being causally responsible for them.'

In developing this theory of manifest objects, I think that I have succeeded in reconciling Hume's vulgar and philosophical systems, but there remains the problem of reconciling this outcome with the contemporary philosophical system, that is to say, the account of the physical world which is presented by contemporary physics. This problem has many ramifications, into which I cannot enter here. Briefly, what seems to have happened is that the perceptible 'constructs' which are needed for the conception of public space are dismissed into the private sector, and their places taken by imperceptible particles. Whether spatial relations can legitimately be severed in this way from their original terms is a debatable question, but it is not obvious to me that they cannot. What we must in any case avoid, as Hume said, is a system which puts physical objects, as they really are, in a duplicate space to which our senses give us no access. If we are to concede physics any pretensions to truth, our interpretation of it must be intelligible.

In the theory of the external world which I have foisted on Hume, personal identity is subordinated to bodily continuity. This was not Hume's own view, but it does no radical violence to his principles. On the contrary, in the section of the *Treatise* which he devotes to this topic, he claims it to be 'evident, the same method of reasoning must be continu'd, which has so successfully explain'd the identity of plants, and animals, and ships and houses, and of all the compounded and changeable productions either of art or nature' (T 259). The difference is just that Hume equates personal identity with the identity of the mind, and defines this without any reference to the body. When I say that he defines it, I allow for his saying that 'the identity, which we ascribe to the mind of man is only a fictitious one,' and like our other ascriptions of identity proceeds from the 'operation of the imagination' (T 259). I am, however, more influenced by the fact that it is vital to Hume's account of the passions, and also to his theory of morals, that one does have a genuine idea of oneself, and he is not so inconsistent as to free this idea from all dependence on impressions. What I therefore take him to mean by calling the identity of our minds 'fictitious' is that it is not what he calls a 'true' identity, that is, the identity of a single unchanging object, but one that can be resolved into a relation between perceptions. There is no implication, any more than in the case of perceptible bodies, that these relations do not really obtain.

There is, indeed, one passage, in the Book 'Of the Passions', in a section in which Hume expatiates on the nature of sympathy, as a factor in the love of fame, where he actually speaks of our having an impression of ourselves. ''Tis evident', he says, 'that the idea or rather impression of ourselves is always intimately with us, and that our consciousness gives us so lively a conception of our own person, that 'tis not possible to imagine, that any thing can in this particular go beyond it' (T 317). If this is not an oversight, the impression must be one of reflection, directed upon the idea of the self, for Hume consistently maintains elsewhere that since no impression is constant and invariable, one has no impression of oneself. He develops this point in a much quoted passage in the first Book of the *Treatise*. 'For my part,' he says, 'when I enter most intimately into what I call myself, I always stumble on some particular perception or other, of heat or cold, light or shade, love or hatred, pain or pleasure. I never can catch myself at any time without a perception, and never can observe anything but the perception.' He makes the apparent concession that some other person 'may, perhaps, perceive something simple and continu'd, which he calls *himself*', but shows this to be ironical by his assertion that 'setting aside some metaphysicians of this kind, I may venture to affirm of the rest of mankind that they are nothing but a bundle or collection of different perceptions, which succeed each other with an inconceivable rapidity, and are in a perpetual flux and movement' (T 252). Hume's readiness to speak for the rest of mankind suggests that his proposition is only disguised as an empirical generalisation. It is rather that he can conceive of nothing that he would count as a pure awareness of oneself.

What then are the relations in virtue of which a bundle of perceptions constitutes a self? Apart from a rather cursory reference to resemblance and causation and the remark that 'memory does not so much produce as *discover* personal identity, by shewing us the relation of cause and effect among our different perceptions' (T 262), Hume barely attempts to give an answer, and in one of the Appendices to the *Treatise*, he admits his failure to find one, and indeed views the whole question of personal identity as posing a problem which he cannot solve. He upholds the negative steps in his argument, that just as 'we have no idea of external substance, distinct from the ideas of particular qualities', so we have no notion of the mind 'distinct from the particular perceptions', a conclusion sustained by the apparent matter of fact that 'When I turn my reflexion on *myself*, I never can perceive this *self* without one or more perceptions; nor can I perceive any thing but the perceptions': that, consequently, it is the composition of perceptions 'which forms the self': that 'all perceptions are distinct': that 'whatever is distinct is distinguishable: and whatever is distinguishable is separable by the thought or imagination', with the result that perceptions 'may be conceived as separately existent, and may exist separately, without any contradiction or absurdity' (T 634–5). But having thus 'loosened' our perceptions he can find no means of binding them together. He says that there are two principles which he cannot 'render consistent', though he believes each of them to be true. They are respectively 'that all our distinct perceptions are distinct existences' and 'that the mind never perceives any real connexion among distinct existences' (T 636).

This statement is puzzling. Evidently, these principles are not inconsistent with one another. What Hume may have meant was that they are collectively inconsistent with the proposition that perceptions may be so 'composed' as to form a self. But even this is not evident. It depends on what we understand by a 'real connexion'. If what is meant is a logical connexion, then there is no reason in logic why perceptions should not be distinct existences, in the sense that we can consistently conceive of their separation, and yet stand, as a matter of fact, in such empirical relations to one another as are sufficient to constitute a self. This is the line taken by William James, who puts forward a Humean theory of the self in his major work, *The Principles of Psychology*, which has remained a classic since its first publication in 1890. The theory relies on what James takes to be the experienced relations of sensible compresence and sensible continuity.

Apart from a possible charge of circularity, the most serious objection to any theory of this type is that we are not continually conscious. We need, therefore, to find some means of linking the perceptions which lie on either side of an interval, say, of dreamless sleep, and there is at least no obvious candidate for this role, except their common relation, the nature of which itself presents a problem, to one and the same body. To be content to say that they belong to the same mind is unhelpful, if it is just a way of saying that they stand in whatever the relation we are looking for happens to be, and unintelligible if it refers them to the same underlying mental subject. As Hume succinctly puts it in his *Abstract* of the *Treatise*, 'the mind is not a substance in which the perceptions inhere' (T 658).

Neither is this the only difficulty. A point which oddly escapes Hume's sceptical notice is that one attributes identity to persons other than oneself, and that these attributions depend upon the identification of their bodies. It is not as if there were a common store of perceptions, which we could sort into the requisite bundles, and then each light upon the particular bundle that constitutes himself.

We are not aware of the experiences of others in the way that we are aware of our own. This does not deter us from applying a psychological criterion of identity to other persons, or even to letting it override the physical criteria. One can imagine circumstances in which one could attach a sense to saying that two or more persons were simultaneously housed in the same body or, as the result perhaps of a brain-transplant, if this were physically possible, that the same person occupied different bodies at different times. Even so, the primary identification, from the point of view of others than the person or persons concerned, would still be that of the body, and if I am right in thinking that self-consciousness implies the discrimination of oneself from other conscious beings, there is a sense in which the factor of bodily continuity remains predominant.

This is not inconsistent with the view that bodies themselves are 'constructed' out of percepts. It is rather a striking illustration of the way in which the theory to which our impressions give rise, on Humean principles, 'takes over' its origins. The impressions are reinterpreted into the theory as states of an observer; and persons figure among the physical objects that the theory, to which we relate our assessments of existence, allows there to be.

4 Cause and effect

No element of Hume's philosophy has had a greater and more lasting influence than his theory of causality. It has been frequently attacked, and frequently misunderstood. Not all the misunderstanding should be put down to the ill-will of Hume's critics. To some extent he courted it, though I agree with F. P. Ramsey, whose own theory, as set out in one of the Last Papers of *The Foundations of Mathematics*, is basically akin to Hume's, that Hume 'gave his readers credit for more intelligence than they display in their literal interpretations'. I shall argue that while Hume is vulnerable on many points of detail, partly because of his misguided insistence on tracing ideas to their origin, and partly because of his tendency to over-simplify the facts, his fundamental tenets not only admit of no answer but thoroughly deserve to carry conviction.

The first point to make clear is that when Hume speaks of 'the relation of cause and effect' he uses the term in a wider and looser sense than is now current. Whereas we are accustomed to distinguish between causal and functional laws, or between causal and statistical laws, or between events which are directly related as cause and effect and those which are related as the effects of a common cause or through their joint derivation from some overriding theory, Hume's usage is such that any law-like connection between matters of fact is characterised as causal. It is true that when he puts probability on a scale of evidence in which the first place is taken by knowledge, defined as 'the assurance arising from the comparison of ideas', and the second by 'proof', where the result of a causal argument is accepted without any of the uncertainty which attends probability (T 124), he goes on to distinguish the probability which is founded on *chance* from that which arises from *causes*; but there is no inconsistency here, since he maintains 'that there must always be a mixture of causes among the chances, in order to be the foundation of any reasoning' (T 126), and the very fact that he makes no provision for statistical laws except as probabilities founded on *causes* confirms the point that I have been making. Hume's usage can indeed be criticised as favouring the neglect of important distinctions, or even the adoption of an unsatisfactory account of probability, but we shall see that its defects do not vitiate the development of his essential argument.

Another unusual feature of Hume's terminology is that he most frequently speaks of the relation of causality as holding between objects, and indeed so defines it, though he is normally represented as taking it to be a relation between events. This emendation does him no disservice, as his references to objects in this connection can easily be rephrased as references to events, and his inclusion of mental elements, like feelings and volitions, among causes and effects is some justification for it. My own opinion is that his intentions are best represented by construing the relation, in his view of it, as holding between matters of fact, into which objects and events, actions and passions, states and processes, whether physical or mental, can be made to enter, according as the example makes one or other choice of terms appear more suitable. This course not only has the advantage of conforming to Hume's enlargement of the scope of the relation but, since matters of fact are correlative to true propositions, it also helps to bring out the point that Hume is concerned with the causal relation primarily as a ground for inference.

This point may be put more strongly. We have seen that Hume takes there to be three relations on which our association of ideas depends, those of resemblance, contiguity, and cause and effect, but there is a functional difference between the first two and the third. As good a way as any of expressing this difference would be to say that whereas the first two provide tracks for the movements of our attention, the third is the main source of supply for our factual beliefs. The impression or idea of an object is apt to arouse the idea of one that has been found to resemble it or has appeared contiguously to it, or contiguously to an object of a similar sort, but the process goes no further. It does not lead to any beliefs, beyond those that already lie within the domain of memory or the senses. Taken singly, or together, these relations do not persuade us of the existence of particular realities that have not yet come within our experience. For this we have to rely upon reasoning and, as Hume puts it in the *Abstract*, and similarly in the *Enquiry*:

'Tis evident, that all reasonings concerning *matters of fact* are founded on the relation of cause and effect, and that we can never infer the existence of one object from another, unless they be connected together, either mediately or immediately. In order therefore to understand these reasonings, we must be perfectly acquainted with the idea of a cause; and in order to that, must look about us to find something that is the cause of another. (T 649)

There is, however, more to the understanding of our reasonings concerning matters of fact than the analysis of the relation on which they are based. There is also the question of the support which the reasonings draw from the relation, and it is in the way it illuminates this second question that the importance of Hume's theory almost wholly lies.

This being said, we can begin with Hume's actual description of causality. His procedure is to distinguish the elements which enter into the common idea of the relation, and then look for the impressions from which they have been derived. Apart from the confusion of psychological with logical questions of which he is generally guilty, his method is open in this instance to a charge of circularity. As we have already remarked, the level at which he views causality in operation is that of physical objects and events, and in attaining this level he already needs to make use of factual inference. This circularity will not, however, appear to be vicious, and while his genetic approach does lead him to give too limited an account of the causal relation, especially in view of the amount of work with which he charges it, the harm which results does not go very deep. His principal theses are not affected by it.

In analysing our idea of the relation of cause and effect, Hume finds it to be compound. It essentially comprises the relations of priority, contiguity and what he calls necessary connexion. This turns out not to be a relation at all, in any straightforward sense of the term, but Hume at least begins by speaking of it as such. In the case of the first two relations the matching of ideas to impressions is allowed to be obvious. What does present a most serious problem for Hume is the discovery of an impression from which the idea of necessary connexion could possibly be derived. Before we examine Hume's handling of this problem, which leads to the nerve of his theory of causality, I should perhaps say that I think it doubtful whether any of the three elements which Hume takes to be essential to our idea of the causal relation really is so. The reason he gives for making contiguity essential is that

> tho' distant objects may sometimes seem productive of each other, they are commonly found upon examination to be link'd by a chain of causes, which are contiguous among themselves and to the distant objects; and where in any particular instance we cannot discover this connexion, we still presume it to exist. (T 75)

This may indeed have been true of the general opinion of Hume's time, but there is nothing contradictory in the notion of action at a distance, and a scientific theory which admitted it would not now be rejected merely on that ground. Not only that, but Hume himself allows causality to operate in a whole class of cases where there is not just an absence of spatial contiguity, but no question of there being any spatial relation at all. These are the cases in which our thoughts and feelings have physical causes and effects. For in Hume's view, only what was coloured or tangible could literally be accorded spatial properties: he queries our title to attribute spatial location to the data of the other senses, even on the basis of their association with the objects of sight and touch, and in the case of thoughts and feelings, he holds that the maxim 'That an object may exist, and yet be no-where', is obviously satisfied. Whether he is right on this point is debatable. I suppose that the popular answer might turn out to be that our thoughts are inside our heads, but unless we take what I still consider to be the unjustified step of identifying mental with cerebral events, this can hardly be taken as literally true. And whatever reasons there may be for giving our thoughts a figur-ative location, say by assigning them to some area of the brain on which they are held to be causally dependent, this still will not yield the result that they are spatially contiguous to their causes.

Hume does have an argument for insisting that causes must precede their effects, in spite of what would appear to be the possibility, in certain cases, of their being contemporaneous.

> 'Tis an established maxim [he says], both in natural and moral phil-osophy, that an object, which exists for any time in its full perfection without producing another, is not its sole cause: but is assisted by

some other principle, which pushes it from its state of inactivity, and makes it exert that energy, of which it was secretly possest. (T 76)

From this Hume infers that if any cause can be 'perfectly con-temporary with its effect' all of them must be. The reasoning is not made fully explicit, but appears to rest on the assumption that every set of sufficient conditions produces its effect as soon as possible, so that if it were possible for a cause, in this sense, to produce a contemporaneous effect, any set of conditions which failed to do so would not be sufficient. If we further assume determinism, as we shall see that Hume does, so that every effect is at least part of a further cause, we do away with succession and so with time. For, as Hume puts it, 'if one cause was con-temporary with its effect, and this effect with *its* effect, and so on ... all objects must be co-existent' (T 76).

Hume appears to have had some doubts about the validity of this argument, but protected himself by adding that 'the affair is of no great importance'. He was right on both counts. Nothing in his further analysis of causality depends on causes' having to precede their effects, and his argument *is* invalid. What its prem-ise entails is that there can be no interval of time between the cause and the effect, but this does not exclude their overlapping, and there is no reason why, if the overlap is partial in some cases, it should not be complete in others. Neither is the premise itself compelling. If, as Hume does, one is going to make the concept of causality dependent on that of law, there is no logical ground for excluding any interval of time between the states of affairs which a law conjoins. We can admit the possibility of action at a distance in time as well as in space.

I said earlier that I doubted whether the element of necessary connexion was essentially comprised in the popular notion of causality. This was partly out of charity, since I am going to argue that unless it is given a very artificial interpretation, such as we shall see that it receives from Hume, the term does not apply at all to matters of fact, and I should prefer to avoid saddling the public with such confusion of thought that its con-cept of causality had no application. Even so, I should not be surprised if a social survey revealed that most people associated with causality some vague idea of power, or force, or agency, and in that case I should preserve charity by severing these ideas from the description of the actual factors which make their causal judgements acceptable.

However this may be, the important point, at this stage of the argument, which Hume acknowledges, is that even if they were necessary, the relations of priority and contiguity are not sufficient for causality. Something more, or perhaps something wholly different, is required. What can it be? Hume offers an answer, but we shall find that the answer itself is of less consequence than the route which takes him to it, and that the splendour of the route owes less to what Hume asserts, as he pursues it, than to what he denies.

The first of his denials is that there is a logical relation between independent matters of fact. As he himself put it in the *Treatise*, we can rely on the principle 'That there is nothing in any object, considered in itself, which can afford us a reason for drawing a conclusion beyond it' (T 139). In the *Enquiry* he offers the further argument that 'Whatever is intelligible, and can be distinctly conceived, implies no contradiction, and can never be proved false by any demonstrative argument or abstract reasoning *a priori*' (E 35). He gives various examples: maintaining, for instance, that he can clearly and distinctly conceive that a body resembling snow in all other respects 'has yet the taste of salt or feeling of fire', that the proposition 'that all the trees will flourish in December and January, and decay in May and June' is perfectly intelligible (E 35); that from the sole premise that a billiard ball is moving in a straight line towards another, it would never be possible 'to infer motion in the second ball from the motion and impulse of the first' (T 650). If in each of these cases, or in the many other examples that he gives, we infer a contrary conclusion, it is because we are making projections from our past experience. So far as logic is concerned, anything may produce anything.

This step is undoubtedly valid, but some care is needed in setting the argument out. We must avoid being misled by the fact that it is common practice to describe objects in terms of their actual or possible relations to one another, and that very often, especially in the case of artefacts, our descriptions of them include a tacit or explicit reference to their causal properties. For instance, in calling something a pen we imply that it is designed to serve the purpose of producing legible marks; a looking-glass does not deserve its name unless it has the capacity of reflecting images; a match is something that, under specifiable conditions, produces flame when it is struck. Not only is there an enormous number of examples of this sort to be found in common usage, but we can take the process to any length we please. In any case in which we want to make the claim that two properties are invariably associated, we can guarantee their connexion by the simple device of reconstruing the predicate which has so far stood for only one of them in such a way that it comes to stand for their conjunction. Similarly, it is often in our power to reconstrue sentences which express empirical generalisations by framing a deductive theory in which they are treated as expressing definitions or their logical consequences.

It is, however, obvious that manœuvres of this sort afford no real protection against Hume's argument. The empirical questions which we apparently suppress by concealing them under definitions come to light again when we ask whether the definitions are satisfied. If we can manipulate predicates so as to establish logical connexions, we can equally reverse the process. A complex property can be resolved into its separate elements; and the predicates involved so construed as to make it an empirical question whether they are satisfied in common.

But may there not be limits to our power of dismantling a given structure of logically connected concepts? Hume's argument requires that every object be, as he puts it, 'consider'd in itself'. Is it quite certain that this is always possible? There does indeed, seem to be no difficulty so long as we are dealing with sensory qualities or even with particular impressions, but the matters of fact to which Hume applies his argument are not at this elementary level. They concern the behaviour of physical objects, and there is a ground for saying that these cannot be treated individually, as if nothing else existed. They are located in a spatio-temporal system, and this may be taken to entail that the identification of any individual object carries the presupposition that there are others to which it stands in some spatio-temporal relation. There is, however, a saving clause in that this reference to other objects in quite general. Within the whole range of objects to which the one that is being identified will be spatio-temporally related, there will be none in particular to which the identification need carry any reference. For this reason we can fulfil Hume's requirement that objects be considered in themselves, by taking it to imply that with regard to any two objects x and y we can identify either by a description which carries no reference to the other, and then, if this condition is satisfied, his conclusion that, from a statement which asserts the existence of x, under such a description, nothing whatsoever can be deduced about the existence or non-existence of y, emerges as the tautology that it anyhow needs to be.

It is well for Hume that he can avail himself of this tautology, since his other argument that the contrary of any matter of fact can be distinctly conceived is not conclusive. Its weakness lies in the assumption that what we find intelligible must be logically possible. A counter-example was put forward by Professor W. C. Kneale in his book *Probability and Induction*, published in 1949. It is generally though not universally held that a proposition of pure mathematics is necessarily true, if it is true at all. So if this is so, and if Hume were right, the contradictory of a true mathematical proposition should not be conceivable. But now consider Goldbach's conjecture that every even number greater than 2 is the sum of two primes. This has never been proved; neither has an exception been discovered. If we allow mathematical propositions to be true or false, independently of our having a proof or disproof of them, and if we insist on their necessity, then we must conclude that either Goldbach's conjecture or its negation is logically false. Yet each seems equally conceivable.

Even if one is not willing to make the assumptions which are needed to sustain this example, I think it should be conceded that the appeal to what we can imagine does not provide us with a foolproof criterion of logical possibility. Not all contradictions are immediately obvious, and conversely propositions which are logically possible, even propositions which are true, may defeat one's imagination. For instance, the curvature of space is something that we have been taught to believe in, but I suppose that before Einstein propounded his Theory of Relativity, most people would have found it inconceivable. I think, therefore, that while Hume's examples of changes in the usual course of nature have some force, it is on our ability to dissever the logical connections which our language forges between our descriptions of matters of fact that, as I said, his argument securely rests.

As it happens, the position which Kneale was trying to defend when he made use of Goldbach's conjecture was not that causal relations exemplify logical entailments, but rather that the laws of nature are what Kneale called Principles of Necessitation. This view is of especial interest, if only because it brings us to the second of Hume's important denials, which was that there could be any such thing as natural necessity, when this is taken to imply that there can be what Kant was to call synthetic relations between matters of fact, which are necessary even though their existence is not logically demonstrable.

Hume's argument is simply that no such relations are observable. If we take his favourite example of a game of billiards, we do indeed speak in 'forceful' terms of the cue being used to 'strike' the ball and of one ball 'cannoning' off another. But all that we actually observe is a series of changes in spatio-temporal relations. First there is a movement of the player's arm, coinciding with a movement of the cue, then an instant at which the cue and the ball are in spatial contact, then a period during which the ball is in motion, relatively to the objects in its neighbourhood, then an instant at which it is in spatial contact with a second ball, then a period in which both balls are in motion, and then, if the player achieves his cannon, an instant at which the first ball is in spatial contact with a third. In the whole of such a process there is no observable relation for which terms like 'power' or 'force' or 'necessary connexion' could be needed to provide a name. And the same applies to any other example that one might have chosen, whether it be a sequence of physical events, or a conjunction of physical properties. As Hume would put it, we are never furnished with an impression from which the idea of necessary connexion could be derived.

If one is looking for such an impression, the most obvious place to seek it is in one's own experience of action. Do we not find it in the exercise of our wills? Hume considers this suggestion in the *Enquiry* and raises three objections to it. The first is that we do not understand the principle of 'the union of soul with body', which we should do if volition gave us the impression of power: the second is that we cannot explain why we move some of our bodily organs and not others, whereas we should not be embarrassed by this question if we were conscious of a force operating only in favoured instances: the third is that 'we learn from anatomy' that strictly speaking we have no power to move our limbs at all, but only to set in motion nerves, or 'animal spirits', from which the motion of our limbs eventually results; and certainly we are not conscious of any relation of force between the exercise of our wills and the movements of these 'animal spirits' or anything like them (E 64–7).

I think that these arguments can be resisted. It has often been claimed that when we perform voluntary movements or handle objects in the various ways that we constantly do, we undergo an experience which fairly answers to the description of being the experience of bringing something about, and the fact that there are limits to what we can do in this way, and that we may be ignorant of the physical conditions which have to be satisfied if we are to accomplish anything of this sort at all, are not in themselves sufficient to make this description inapt. Indeed Hume himself speaks, in cases where we find ourselves trying to overcome physical resistance, of our experiencing an 'animal nisus', and acknowledges that it enters very much into the vulgar idea of power, even if this vulgar idea is neither precise nor accurate (E 67).

But now I must say that, whatever the rights or wrongs of this question may be, and however great its psychological interest, it is only peripheral to Hume's argument. What we must remember is that Hume is concerned with causality as the ground of factual inference; it has to supply the bridge which carries us safely from a true belief in one matter of fact to a true belief in another. And the consequence of this is that even if we do have experiences which are properly described as experiences of the exercise of power, they are nothing to the purpose. The reason for this is that they cannot be generalised. From the fact, if it be a fact, that a faithful description of some particular action of mine implies, either overtly or covertly, that I have undergone such an experience, it in no way follows that when I set myself to repeat the action, I shall have the same experience or indeed achieve the same result. There being no logical connection between the two performances, we can make our description of either one of them as elaborate as our psychological theories permit; it will still convey nothing about the nature of the other.

The same applies to the search for necessity in the physical world. Here again, Hume had failed to do full justice to his case by making it appear to rest upon an empirical generalisation, rather than a logical argument. If we return to his example of the billiard balls, I think that he is right in saying that no relations of force or power can be detected in the actual phenomena. The point which he overlooks is that it would not matter if they were detectable. For let us suppose that when the balls collided we did observe something which could fairly be described as the communication of force, and that we therefore included a reference to it in our description of the facts. So far as our right to make inferences goes, this complication leaves us exactly where we were. Let us call the two balls 'A' and 'B' and the supposedly forceful relation between them 'R'. Then, if there is any doubt whether the sequel to the spatial coincidence of A and B will be the same on a future occasion, there must be an equal, or indeed a stronger doubt, since more is being assumed, whether A will again stand in the relation R to B on a future occasion of their spatial coincidence, and whether there will be the same sequel. It may be objected that the doubt is settled in the second case by the very nature of the relation R. If A stands in relation R to B, it *must* impart motion to it, and this implies that under similar conditions it does so on all occasions. But this objection is simply fallacious. For either R is a merely phenomenal relation, that is to say, a relation such that an accurate observation of the phenomena is sufficient to establish its presence, or it is not. If it is, its existence will be entirely neutral with regard to what happens at any other place or time. If it is not, it can serve its intended purpose only if it comprises in its definition the clause that any terms which are related by it will always, under similar conditions, exhibit similar behaviour. But this makes the example just another case in which causal propositions are made true by fiat, and the same objections are fatal to it. Moreover, on this interpretation, what looked like the empirical proposition that necessary connexion is not detectible in any single instance of a conjunction of matters of fact, is promoted into a logical truth. For the relation is now defined in such a way that we have to examine every single instance of the phenomenon in question, in order to discover that it holds in any one of them.

There the matter of necessity might be allowed to rest, though much more would remain to be said about our grounds for making factual inferences. But assuming as he does that we have an idea of necessary connexion, Hume is driven by his principles to continue his search for the impression which could be the source of this idea. He sees that the mere multiplication of instances will not supply it. Having stated the principle, which we have already quoted, '*That there is nothing in any object, consider'd in itself, which can afford us a reason for drawing a conclusion beyond it*', he goes on to state the further principle, '*That even after the observation of the frequent or constant conjunction of objects, we have no reason to draw any inference concerning any object beyond those of which we have had experience*' (T 139), and if we are restricting ourselves, as Hume here does, to deductive inference, the truth of the second of these principles is no less evident than that of the first. Nevertheless it is in the multiplication of instances that Hume finds the clue that leads him to the end of his quest. His theory is that the observation of the frequent or constant conjunction of matters of fact of recurring types gives rise to a mental habit or custom of expecting this regularity to be repeated. The difference made by the multiplication of instances is, as Hume puts it in the *Enquiry*, that 'the mind is carried by habit, upon the appearance of one event, to expect its usual attendant, and to believe that it will exist' (E 75). It is, then, in 'this connexion which we *feel* in the mind, this customary transition of the imagination from one object to its usual attendant' that Hume discovers 'the sentiment or impression from which we form the idea of power or necessary connexion' (E 75). All that we have to go on is our past experience of regularity in nature; indeed there is nothing more there to be discovered. We become habituated to expecting that this regularity will be maintained. And this habit or custom is so firmly ingrained in us that we project the strength of the association into the phenomena themselves, and so succumb to the illusion that 'necessary connexion' is the name of a relation in which they actually stand to one another.

This leaves the way clear for Hume to define causality. Both in the *Treatise* and in the *Enquiry* he offers alternative definitions, according as the relation is considered as 'natural' or 'philosophical': that is to say, according as we attend only to the phenomena in which the relation is exhibited, or also take account of the light in which we view them. In the first case, a cause is defined in the *Treatise* as 'An object precedent and contiguous to another, and where all the objects resembling the former are plac'd in like relations of precedence and contiguity to those objects, that resemble the latter'. The alternative, 'philosophical' definition is that 'A cause is an object precedent and contiguous to another, and so united with it, that the idea of the one determines the mind to form the idea of the other, and the impression of the one to form a more lively idea of the other' (T 170). The definitions given in the *Enquiry* are similar but more succinct. In its natural aspect a cause is said to be 'An object followed by another, and where all the objects similar to the first are followed by objects similar to the second', with the additional gloss, 'where, if the first object had not been, the second had never existed': when the contribution of the mind is brought in, a cause becomes 'an object followed by another, and whose appearance always conveys the thought to that other' (E 76–7).

It has often been pointed out, and hardly needs repeating, that these definitions are far from being adequate, as they stand. Apart from the infelicity, which we have already noted, of speaking of 'objects' as the terms of causal relations, and the unwarranted exclusion of action at a distance, they suffer from their failure to take account of the part played by theories in the derivation of what we take to be causal laws. It is doubtful, also, whether the conjunction need be entirely constant. We often make particular causal judgements, with no stronger general backing than a statement of tendency. Hume, indeed, makes some provision for cases of this sort when he speaks of 'the probability of causes'. He assumes, not altogether in accord with common usage, that a cause has to be a sufficient condition, and also unduly takes it for granted that there is never more than one sufficient condition for any given matter of fact. This would make a cause also a necessary condition, which is, indeed, what Hume describes it as being in the gloss to the first of his definitions in the *Enquiry*. He allows, however, for what he calls the contrariety of causes, which means that the presence of some other factor, or factors, prevents a given type of 'object' from being followed by its usual attendant. If we knew what these other factors were and how they operated we could make their absence part of our sufficient condition and regulate our expectations accordingly. As it is, we have often to be content with generalising from past frequencies. This account of our procedure is not objectionable in itself, so long as we admit the assumption of there not being a plurality of sufficient conditions. What it overlooks is the caveats which we impose on statistical inferences, and once more the extent to which statistical laws are derived from theories.

Conversely, not every constant conjunction is seen as warranting an ascription of causality. In cases where the examples are not numerous, or occur in what are thought to be exceptional circumstances, or do not fit in with our general picture of the way things happen, the conjunction may be deemed accidental, even though it has occurred without exception. This goes with the serious objection that Hume's use of words like 'similar' and 'resembling' in his definitions is too vague. Any two objects may be similar in some respect or other. We need to be told in some detail what kind or degree of similarity is required for the matters of fact which it groups to become suitable candidates for factual inference.

The charge of circularity is sometimes brought against the second of Hume's definitions in the *Treatise*, on the ground that he speaks of an object as 'determining' the mind to form the idea of another. This charge is unjustified. It is clear from the whole course of Hume's argument that no more is claimed here than that the mind in fact acquires the habit of associating the ideas in question. There is no implication that it is 'forced' to do so. On the other hand, it was a venial mistake on Hume's part to include a reference to the mind's propensity in what was supposed to be a definition of causality. In propounding causal judgements, we express our mental habits, but do not normally assert that we have them. An account of our mental habits does enter into the explanation of our ascriptions of causality: but this is not to say that when we attribute causal properties to some physical object, we are also making an assertion about ourselves.

Enough has been said to show that Hume's definitions are formally defective. The fact remains that they do bring out two points, the importance of which outweighs their defects. The first is that it is only the existence of the appropriate regularity in nature that can make a causal proposition true: and the second is that the difference between accidental and causal generalisations is not a difference in the ways they are satisfied, but a difference in our respective attitudes towards them. In the second class of cases we are willing to project accredited regularities on to imaginary or unknown instances; in the first we are not. Though he points the way to this distinction, Hume does not himself enquire into the principles which underlie it.

Instead, he raises the more general and fundamental question how we can ever be justified in making factual inferences which carry us beyond the evidence of our past and present observations. In raising this question, he posed what has come to be known to philosophers as the Problem of Induction. He approaches it by asking whether our procedure is governed by reason. If it were, he maintains that our reason 'would proceed upon that principle, *that instances of which we have had no experience, must resemble those, of which we have had experience, and that the course of nature continues always uniformly the same*' (T 89). But now we come to the next of his crucial denials. For he shows convincingly that the principle at issue can neither be demonstrated nor even lay claim to any probability. That the principle cannot be demonstrated follows evidently from the assumption, which we have seen that Hume is entitled to make, that the members of the two classes of instances are logically distinct. As for its being probable, we run up against the fact that the ascription of probability draws its force from past experience. In Hume's words, 'probability is founded on the presumption of a resemblance betwixt those objects, of which we have had experience, and those, of which we have had none; and therefore 'tis impossible this presumption can arise from probability' (T 90).

Does it follow that we have no good reason at all for trusting the outcome of any factual inference? From Hume's point of view, this would be a misleading way of stating what he has proved, since it suggests that we lack the good reasons we might conceivably have had. His own conclusion is rather that which Kemp Smith attributes to him: that reason has in such matters been ruled out of court. In a section of the *Treatise* in which he allows scepticism to invade the domain of reason, using the argument that the admitted possibility of our making mistakes in the demonstrative sciences entails that even in their case knowledge degenerates into probability, with the surely fallacious rider that, the judgement of probability itself not being certain, the doubts successively mount up until even the probability is all taken away, he remarks that the question whether he himself accepts the totally sceptical conclusion, 'that our judgement is not in *any* thing possest of *any* measure of truth and falsehood', is 'entirely superfluous', since it is not an opinion that anyone can sincerely hold. If he has developed the sceptic's case to its greatest extent, it is only to obtain evidence for his hypothesis *'that all our reasonings concerning matters of fact are deriv'd from nothing but custom: and that belief is more properly an act of the sensitive, than of the cogitative part of our natures'* (T 183). This echoes an earlier statement: 'Thus all probable reasoning is nothing but a species of sensation. 'Tis not only in poetry and music, we must follow our taste and sentiment, but likewise in philosophy' (T 103).

Does this mean that all factual reasoning is on the same footing; that we are entitled to extrapolate from our past and present experience in any way that suits our fancy? If Hume really thought so, it is strange that he should have applied what he called 'the experimental method' to the study of the passions or that he should have laid down a set of 'Rules by which to judge of Causes and Effects' (T 173ff.), such as that there must be a constant union between them, and 'that the difference in the effects of two resembling objects must proceed from that particular, in which they differ'. Why is he so sure that 'there can be no such thing as Chance in the world' (E 56), so that when we ascribe an event to chance we are confessing our ignorance of its real cause, a question made all the more puzzling by his having been at considerable pains to show (T 78ff.) that the generally received maxim that *'whatever begins to exist, must have a cause of existence'* is neither intuitively nor demonstrably certain, and to expose the fallacies in the arguments by which various philosophers had sought to prove it? The only explanation seems to be that the propositions that 'every event has a cause' and that 'the course of nature continues always uniformly the same' were regarded by Hume in the light of natural beliefs. They cannot be proved, but nature is so constituted that we cannot avoid accepting them.

But can we really not avoid it? It seems to me that we can, in both instances, though I shall not here dwell upon the first. The trouble with the second proposition is that it is not clear how it is to be construed. If it is taken at its face value, as stating that the course of nature is entirely repetitious, then so far from its expressing a natural belief, it is very unlikely that anybody believes it. Our experience leads us to expect the occurrence of unforeseen events; at the most we may believe that they can subsequently be accounted for. On the other hand, if no more is being claimed than that past experience is, in some large measure, a reliable guide to the future, there is no question but that this is universally assumed. There still, however, remains the problem of explaining why not every instance of experienced conjunction is thought to be equally projectible.

This goes with the fact, which has been strangely overlooked by the many philosophers, starting with Kant, who have tried to rebut Hume by arguing for the necessity, or at least the probability, of some general principle of uniformity, that its very generality would prevent such a principle from doing the work required of it. I am not sure whether Hume believed that the adoption of the principle which he formulated would legitimise inductive arguments by making them deductive, but if he did, he was mistaken. This can easily be shown by considering any case in which a universal generalisation is inferred from an unexhausted series of wholly favourable instances. Suppose that a number of objects of the kind A have been observed and all found to have the property f. Then the suggestion is that by adding the premise that nature is uniform we can deduce that every A has f. But it still may happen that we come upon an A which lacks f. In that case we shall not only have suffered the not unprecedented setback of discovering that a generalisation which we had accepted is false. We shall have proved that nature is not uniform, since if the conclusion of a valid deductive argument is false, at least one of its premises must be false, and in view of the truth of the premise that all observed As have been *f*, the false premise has to be the principle of uniformity. But surely neither Hume, nor any other philosopher who has thought there to be a need for such a principle, can have seriously intended it to be so stringent that it succumbed to the discovery of a first exception to any generalisation for which there had been any positive support.

Let it be granted, then, that there is no such simple method of transforming inductive into deductive reasoning. It may still be held that the use of a general principle of uniformity, when combined with the evidence of previous observations, can serve to bestow a high measure of probability on some generalisations and still more upon our expectations of particular events. This was very much the line taken by John Stuart Mill, though it has been shown that he needed to make further assumptions for his methods to yield the results that he claimed for them. His position was, however, avowedly circular. The principle of uniformity was itself viewed as obtaining its support from the generalisations which it joined in supporting.

It has been suggested that the circle can be avoided by deriving a set of general principles from a combination of the deliverances of past experience with *a priori* relations of probability, themselves based on the mathematical calculus of chances. The short answer to this is that no such relations are available, for the reason which Hume already gave in his discussion 'of the probability of chances'.

Shou'd it be said [he remarks], that tho' in an opposition of chances 'tis impossible to determine with *certainty*, on which side the event will fall, yet we can pronounce with certainty, that 'tis more likely and probable, 'twill be on that side where there is a superior number of chances, than where there is an inferior: Shou'd this be said, I wou'd ask, what is here meant by *likelihood* and *probability*? The likelihood and probability of chances is a superior number of equal chances; and consequently when we say 'tis likely the event will fall on the side, which is superior, rather than on the inferior, we do no more than affirm, that where there is a superior number of chances there is actually a superior, and where there is an inferior there is an inferior; which are identical propositions and of no consequence. (T 127)

In short, the mathematical calculus is a purely formal system, and if we are to apply it to our estimates of what is actually likely to happen, we need to make some such empirical assumption as that where we have a set of mutually exclusive possibilities, like the coming uppermost of one or other of the six faces of a die, and no information which favours one outcome rather than any other, they can be expected to occur with approximately equal frequency. But then we are back in our circle, for we have no reason to make any assumption of this sort except on the basis of past experience. Mere ignorance does not establish any probability.

There are those who maintain that Hume, and his followers and opponents, have all been tilting at windmills, since scientists do not employ inductive reasoning. They advance hypotheses, submit them to the severest tests that they can devise, and adhere to them so long as they are not falsified. I doubt, for my own part, if this is a fully accurate account of scientific procedure, though it serves as a corrective to the mistaken view that the practice of science consists in nothing more than generalising from past observations. But in any case it does not entail that induction is or could be superfluous. For one thing, this suggestion overlooks the fact that an enormous amount of inductive reasoning is built into our language. In referring to physical objects of whatever sort, we are implying that properties which have hitherto been found in conjunction will continue to be so; in crediting these objects with causal powers, we are forecasting the repetition, under suitable conditions, of previous sequences of different sorts of events. Furthermore, there would seem to be no point in testing hypotheses unless their passing the test was thought to enhance their credibility: but that it does so enhance it is an inductive assumption.

A more promising line of argument was advanced long ago by C. S. Peirce. If there is no ascertainable generalisation to which a given set of facts conforms, there is nothing we can do about it; if there is such a generalisation then the reliance on past experience will eventually lead us to it. There is, however, the difficulty that there may be an infinite number of generalisations with which any limited set of observations might be consistent, and there is not much comfort in the assurance that we are bound to get the right result in an infinite time.

Even so, there is some point in remarking that any successful *method* of forming our expectations must be inductive, for the sufficient reason that it would not be a successful method unless it followed a pattern which corresponded to the pattern of events with which it dealt. The real trouble lies in the latitude which our observations allow us, so that we have to choose between a number of competing hypotheses which are equally well supported by our past experience. The question at issue is not so much *whether* the future will resemble the past, since if the world is to continue to be describable at all, it must resemble it in some way or other, but *how* it will resemble it. What we want and cannot obtain, except by circular argument, is a justification for our actual interpretation of the lessons of the past; a justification for adhering to a special corpus of beliefs. That we cannot obtain it is an insight which we owe to Hume. What he neither proved nor even sought to prove was that the consequence is that the beliefs should be abandoned.

5 Morals, politics and religion

Considering the relish with which Hume denies the necessity of the proposition that every event has a cause, we may be surprised to discover how firmly he accepts its truth. There is, indeed, a passage in the *Treatise* where, in the course of discussing the passions of fear and hope, he speaks of the probabilities which cause them as belonging to one or other of two kinds, according as the object is already certain but 'uncertain to our judgement' or 'when the object is really in itself uncertain, and to be determin'd by chance' (T 444), but this is a rare departure from his usual view that 'what the vulgar call chance is nothing but a secret and conceal'd cause' (T 130). Throughout his analysis of the passions, and his exploration of the foundations of morals and politics, his official position is that of a determinist.

In speaking of Hume as a determinist, we must, however, bear in mind that this does not in his case carry any pledge of allegiance to a reign of necessity in nature. The most to which it commits him, or any other determinist who understands what he is talking about, is the proposition that the concurrence and succession of matters of fact of different types exhibit perfect regularity. In accepting this proposition, Hume not only sets aside the doubts which he has raised about the validity of induction, but also discounts the fact that the regularities which our past experience has brought to light are far from being perfect. They do not cover all the accredited data and their level of generality allows for some unexplained latitude among the instances in which they are exemplified. Hume takes no notice of this second point, but he does acknowledge the first. He disposes of it by contrasting the opinion of 'the vulgar, who take things according to their first appearance', with that of philosophers who attribute any appearance of irregularity to 'the secret opposition of contrary causes' (E 86–7). The objection to this course is that it reduces the thesis of determinism to a convention. It survives in all circumstances if we are allowed to set aside any adverse evidence by an appeal to secret powers. To endow it with some interest, we need to break it down into a set of working theories which we then apply to different domains of fact. It is true that if our theories fail, we may retain the hope of finding replacements which will bring us success, but at least at any given stage the questions which we are asking will be empirical.

Though these considerations did not occur to Hume, there is no reason to think that he would have found them unwelcome. For the proposition which he wished to advance was not so much that the operations of bodies are strictly regulated as that there is no distinction in this respect between body and mind. The character of human nature is as constant as that of inanimate objects and what is more 'this regular conjunction has been universally acknowledged among mankind, and has never been the subject of dispute, either in philosophy or common life' (E 88). If we are startled to find this said of philosophy, we need to remind ourselves that the term, in Hume's usage of it, encompasses science; and what he is here maintaining is that the custom of drawing inferences from past to future experience, as well as the assumption of firmer regularities than we have actually discovered, occur no less in the social than in the natural sciences.

What [he asks] would become of *history*, had we not a dependence on the veracity of the historian, according to the experience which we have had of mankind? How could *politics* be a science, if laws and forms of government had not a uniform influence upon society? Where would be the foundation of *morals*, if particular characters had no certain or determinate powers to produce particular sentiments, and if these sentiments had no constant operation on actions? And with what pretence could we employ our *criticism* upon any poet or polite author, if we could not pronounce the conduct and sentiments of his actors either natural or unnatural to such characters, and in such circumstances? It seems almost impossible, therefore, to engage either in science or action of any kind without acknowledging the doctrine of necessity, and this *inference* from motives to voluntary actions, from characters to conduct. (E 90)

The conclusion which Hume drew from this is that no one seriously believes in the freedom of the will, if this is taken to imply that men's actions are uncaused. He admits that when it comes to one's own actions one has a tendency to claim this type of freedom. 'We feel, that our actions are subject to our will, on most occasions; and imagine that our will itself is subject to nothing' (E 94), partly because we do not suffer from any feeling of constraint. We may be ignorant of all the regularities to which we conform, and even when we discover them we may still fancy that we can elude their grasp. We may even prove this to our own satisfaction by choosing a different course of action in what we take to be the same circumstances, not seeing that there is a decisive difference in that

> the fantastical desire of shewing liberty is now itself operating as a cause. What is more to the purpose is that, however we may imagine we feel a liberty within ourselves, a spectator can commonly infer our actions from our motives and character; and even where he cannot, he concludes, in general, that he might, were he perfectly acquainted with every circumstance of our situation and temper, and the most secret springs of our complexion and disposition. (E 94)

Since we are intended to side with the spectator, this last claim evidently begs the question, and to the extent that it does so, Hume fails to prove his point. But whatever other moral may be drawn from this failure, it does not strengthen the position of those who wish to maintain the usual association of free will with responsibility. From their point of view, as well as Hume's, the possibility of an overall reign of purely physical causes is ruled out of court. Let us suppose then that our actions do not proceed, with any high degree of regularity, from our character and motives. To what do we ascribe their waywardness? I see no other candidate but chance. But why should we be held responsible for actions which occur by chance, any more than for those that issue from our motives and characters, which may perhaps themselves be explained in terms of our genetic endowment, and the physical and mental stimuli to which we have responded from our infancy?

This is not to deny us any form of liberty. Hume defines it as 'a power of acting or not acting, according to the determinations of the will' (E 95), and this is something that we commonly possess. Not only that, but Hume is able to argue that it is liberty of this sort, rather than the freedom of the will to which we cannot properly lay claim, that is 'essential to morality'. His argument is that 'actions are objects of our moral sentiment, so far only as they are indications of the internal character, passions, and affections' (E 99). They cannot give rise to praise or blame when they are 'derived only from external violence', and equally there would be no point in the distribution of praise or blame, reward or punishment, unless it had an effect upon men's motives and characters, and consequently upon their actions. This does not debar us from praising or blaming the conduct of the dead, or that of persons whom for other reasons we have no actual power to influence, but this is chiefly another illustration of the force of custom. There being certain sorts of conduct that we find it possible and useful, in many cases, to promote or inhibit by the expression of praise or blame, we fall into the habit of reacting in the same way to all conduct of these sorts. It is also to be noted that even when the agents in question are beyond our reach, our evaluations of their actions may still have an effect on the behaviour of those who are inclined to follow their example.

On the question of free will, I think that Hume is right, except for his claim that his definition of liberty is one 'on which all men agree'. It would, I think, be generally agreed that he has laid down a necessary condition of liberty. I doubt, however, if it is commonly thought to be sufficient. It appears to me, rather, that not only our moral judgements but also many of our feelings about ourselves and other persons, such feelings as those of pride or gratitude, are partly governed by a notion of desert, which requires our wills to be free in a stronger sense than Hume's definition admits. In a muddled fashion, we credit ourselves and others with what has sometimes been described as a power of self-determination. The trouble is that even if there were something which answered to this description, we should not escape from Hume's dilemma. Either the exercise of this power would fit into a causal pattern, or it would occur at random, and in neither case would it appear to justify an ascription of responsibility. We could avoid the muddle, as in effect Hume proposes, by altering our notions of desert and responsibility so that they fitted into a purely utilitarian scheme, but whether we have such command over our sentiments as to be able to abide by such a policy, and whether it is wholly desirable that we should abide by it even if we could, are questions which are open to dispute.

A point of interest, which Hume slyly makes in passing, is that if determinism were valid and one assumed the existence of an omnipotent creator, another dilemma would arise for those who wished to maintain that the creator was benevolent. As Hume puts it, 'human actions ... either can have no moral turpitude at all, as proceeding from so good a cause; or if they have any turpitude, they must involve our Creator in the same guilt, while he is acknowledged to be their ultimate cause and author' (E 100). One can escape through the horns of this dilemma by discarding the concept of guilt altogether, which it is arguable that we should do anyway in view of its dependence on the muddled concept of free will, but the escape is quickly followed by recapture, since the dilemma can be restated in terms of evil, with an even heavier burden placed on the creator, since not all the evil in the world results from deliberate human actions. In this case there is no escape, since evil can be taken as comprising a large amount of human and animal suffering, if nothing more, and it is not to be denied that such suffering occurs. It may be alleviated in some cases by the belief that it proceeds from a good cause, but quite apart from the fact that it is not clear how this could achieve any more than a diminution of the evil, such cases are exceptional. Having already rejected the preposterous suggestion that this is the best of all possible worlds, we are left with the conclusion that if there is an omnipotent creator, he is not benevolent.

Characteristically, Hume leaves it to the reader to draw this conclusion. Having presented his argument, he is content to remark that 'to defend absolute decrees, and yet free the Deity from being the author of sin, has been found hitherto to exceed all the power of philosophy' (E 103). Since the weakest link in Hume's argument is his assumption of determinism, it is worth remarking that its conclusion would scarcely be impaired if this assumption were relaxed. The moral standing of a creator would not be very much higher if he had willed only a considerable proportion of the world's evil and left the remainder to chance.

After making his ironical comment that the question of the Deity's involvement in sin has been found to exceed the power of philosophy, Hume recommends that the subject 'return, with suitable modesty, to her true and proper province, the exam-ination of common life' (E 103), and this is the policy which he himself pursues in giving his philosophical account of human morality. Hume's moral philosophy, reviewed in all its detail, is subtle and complex, but it is derived from a small number of principles which can be clearly elicited. They are partly analytical and partly psychological. I shall first state and discuss them, and then consider the conclusions which Hume draws from them.

The principles are the following. No special importance should be attached to the order in which I have set them out.

1 Reason alone, being concerned only with the discovery of truth and falsehood, 'can never be the motive for any action of the will' (T 413). It is from this principle that Hume derives his celebrated dictum: 'Reason is and ought only to be the slave of the passions, and can never pretend to any other office than to serve and obey them' (T 415).

2 The passions by which we are motivated may be direct or indirect, calm or violent. The direct passions, such as those of joy, grief, hope or fear, arise either from natural instinct, or from our desire of good, which can here be equated with pleasure, or aversion from evil, which can here be equated with pain. The indirect passions, such as those of pride, humility, love or hatred, arise from a combination of these primitive motives with other factors. This distinction is independent of that between calmness and violence. It is because the motivation may be violent that 'men often act knowingly against their interest' and are not always influenced by 'their view of the greatest possible good' (T 418).

3 Sympathy for other creatures is a natural instinct. Its strength is such that although it is 'rare to meet with one, who loves any single person better than himself', it is equally 'rare to meet with one, in whom all the kind affections, taken together, do not over-balance all the selfish' (T 487). This natural instinct of sympathy or benevolence plays a large part in the formation of our moral and political attitudes.

4 'Since morals ... have an influence on the actions and affections, it follows that they cannot be deriv'd from reason' (T 457). Accordingly, 'the rules of morality are not conclusions of our reason'.

5 Moral judgements are not descriptions of matters of fact. 'Take any action allow'd to be vicious: Wilful murder, for in-

stance. Examine it in all lights, and see if you can find that matter of fact, or real existence which you call *vice*. In whichever way you take it, you find only certain passions, motives, volitions and thoughts. There is no other matter of fact in the case' (T 468). Similarly, when 'instead of the usual copulations of propositions, *is*, and *is not*', one suddenly meets 'with no proposition that is not connected with an *ought*, or an *ought not*' (T 469), one is being tricked. It is not possible that 'this new relation can be a deduction from others, which are entirely different from it'.

6 'Vice and virtue may be compared to sounds, colours, heat and cold, which according to modern philosophy, are not qualities in the object, but perceptions in the mind' (T 469). Accordingly, 'when you pronounce any character to be vicious, you mean nothing, but that from the constitution of your nature, you have a feeling or sentiment of blame from the contemplation of it'. It is interesting to note that here and elsewhere in his writing about morals, Hume takes the Lockean view of secondary qualities, in spite of his previous rejection of it when he was writing about the understanding.

7 Though one speaks of virtuous or vicious actions, they derive their merit or demerit only from virtuous or vicious motives, and it only as signs of such motives, or the character of the person who acts from them, that actions are subject to moral evaluation.

8 The sentiment of approbation, aroused by 'a mental action or quality' which we therefore call virtuous, is itself pleasing, and that of disapprobation, which is similarly related to vice, displeasing (E 289). We can therefore also take virtue to be 'the power of producing love or pride' and vice to be 'the power of producing humility or hatred' (T 575).

9 What arouses our approbation or disapprobation is the appraisal of qualities or motives as being respectively productive of a preponderance of pleasure or pain. These appraisals may also be characterised as judgements of utility.

10 'No action can be virtuous or morally good unless there is in human nature some motive to produce it, distinct from the sense of its morality' (T 479).

11 The sense of justice, on which both moral and political

obligation depend, is derived not from any natural impressions of reflection but from impressions due to 'artifice and human conventions' (T 496).

Let us begin by examining the psychological principles. There may be some doubt whether the cases in which the kind affections do not overbalance the selfish are so rare as Hume supposes, but I can find no good reason to doubt that we do have a natural instinct of sympathy or benevolence. Attempts have been made to subordinate sympathy to self-love, but they appear to me perverse. Since there can be no reason *a priori* for assuming that every action has a selfish base, it is simpler and more sensible in this instance to accept what Hume calls 'the obvious appearance of things' (E 298).

Without entering into the niceties of Hume's distinctions between the direct and indirect or the calm and violent passions, I think that the following points should be accepted. First, that not all intentional action is motivated, in the sense that it proceeds from a desire to bring about some good either to oneself, or to others, or to society at large, and, secondly, that even when our actions are motivated in this way they may not conform to utilitarian principles. That they rarely conform to purely egoistic principles has already been conceded, but neither are they simply altruistic. Sympathy varies in degree, and its strength depends on a variety of other relations in which one may stand to its object; it is not simply proportionate to one's estimate of the object's worth or need. Thus, one may choose to confer benefit on a member of one's family, on a lover or a friend, at the cost not only of one's own interest but of what one believes to be the greater interest of a wider circle. In the cases where one's actions are not purposive, such emotions as those of pity or embarrassment or anger may lead us to behave, even consciously, in ways that do no good to any one at all. It might be argued that actions done out of pity or anger were purposive, since they implied a desire to do good or harm to the persons concerned, but the answer to this is that the desire proceeds from the emotions, not the other way around. Neither need it be the case, as Hume seems to imply, that actions in which one disregards the consequences proceed only from the violent passions. Very often they are due to inertia. One simply cannot be bothered to realise one's preferences, whether the preference consists in the procurement of pleasure or in getting out of a situation which one finds disagreeable. Again it can be argued that the disvalue which one attaches to making the effort is judged to outweigh the value which one expects to result from its outcome, but unless this account is made trivial by allowing the judgement to be unconscious and then equating inertia with it, I think that it is simply untrue to the facts.

To say that our motivated actions do not in general conform to utilitarian principles is not to deny that they should. It can still be maintained that it is only when they do so that we count them as virtuous. There is, indeed, the further psychological obstacle that not only do we not always do what we want, but even when we are doing what we want, our objectives are commonly more specific then the production of a state of pleasure or the abolition of a state of pain, though here again it could be held that it is only when they have these objectives that our actions are accounted virtuous. I think, however, that it would be a mistake to attribute this view to Hume. Admittedly, he says that 'every quality of the mind is denominated virtuous, which gives pleasure by the mere survey: as every quality, which produces pain, is called vicious' (T 591), adding that 'this pleasure and this pain may arise from four different sources', since 'we reap a pleasure from the view of a character, which is naturally fitted to be useful to others, or to the person himself, or which is agreeable to others, or to the person himself'. There are, however, two points which diminish the force of this quotation. One is that Hume does not think that the pleasure which we take in surveying a virtuous quality is always of the same kind; it varies according to the nature of the quality in question. The other is that when he speaks of a character as being fitted to be useful to others, he does not equate this utility with the maximisation of pleasure. It can consist in bringing about the satisfaction of whatever desires the persons affected happen to have.

In this context, it is also worth remarking that Hume is not a forerunner of utilitarians like Bentham and Mill. We shall see that he associates the conventional virtue of justice with regard to the public interest, but he by no means takes it to be a general feature of the objects of our moral approbation that they promote anything of the order of the greatest happiness of the greatest number.

It has been objected to Hume that there is no such thing as a moral sense, analogous to the physical senses, or a special feeling of moral approbation, which the view of virtuous qualities or characters invariably arouses. The answer to this is that he is not committed to saying that there is. Admittedly, as we have seen, he speaks of our having a feeling or sentiment of blame when we contemplate a character which we pronounce to be vicious, but the point which he is concerned to make is that in calling a character vicious we are not assigning it a special intrinsic property but expressing our reaction to the properties which it has. This reaction must, indeed, be adverse, but it need not take exactly the same form in every case. We do in fact have feelings of revulsion or moral indignation, but they need not always be present in every instance of moral condemnation. There is a large variety of states, dispositions and actions, not all of them infused with emotion, in which our attitudes of moral approval or disapproval may actually consist.

I believe also that some critics have been misled by Hume's use of the word 'mean' when he speaks of our meaning nothing by the use of moral predicates but that the contemplation of the actions or characters to which we apply them causes us to have favourable or hostile feelings. I do not think that Hume is advancing the thesis that a statement of the form 'X is good' is logically equivalent to the statement 'I have a feeling of moral approval in contemplating X' or yet to some such statement as 'The contemplation of X arouses a feeling of approval in most normal men' or 'in most of the members of such and such a society'. And the same applies to statements about the rightness of actions or the obligation to perform them. I do not credit Hume with the view that in making statements of either of these sorts we are covertly asserting something about ourselves or about other actual or possible critics. There is indeed a sense in which he is offering an analysis of our moral judgements, but the analysis is not intended to supply us with a recipe for translating the sentences which express them. It consists rather in an account of the circumstances in which we are induced to employ moral predicates, and of the purposes which their employment serves. Moreover, if we did insist on extracting from Hume a reformulation of our moral statements, we should come nearer the mark by crediting him with the modern 'emotive' theory that they serve to express our moral sentiments rather than with the theory that they are statements of fact about one's own or other people's mental condition.

It is not even clear that Hume is correctly described as a moral subjectivist, unless we are using the term 'subjectivist' in such a way that it is correct to say of Locke that he gave a subjectivist account of qualities like colour. But is it not one of Hume's principles that moral judgements are not descriptions of matters of fact? This is true, but what it means is that they lack the sort of factual content that is to be found in a description of a person's motives or in an account of what he actually did. From another point of view, it is as much a matter of fact that a motive or character of a given sort has the tendency to evoke certain responses in those who contemplate it, as it is a matter of fact that the responses are or tend to be produced. Neither is there any evidence in the texts that Hume wanted to deny this. What he did want to deny was that moral predicates stand for what Locke called primary qualities, or, in other words, that they stand for intrinsic features of the motives or characters or actions to which they are applied.

In this he is surely right. To take his own example of a case of wilful murder, the badness of the act is not an additional feature of it, ranking alongside the facts that the killer had such and such motives, or that he brought about the death of his victim in such and such a fashion. Neither is it manifested as a sort of glaze with which the conjunction of such facts is overlaid. My statement that it was a wrongful act may carry a descriptive content, if it is construed as presupposing my acceptance of some prevalent code of morals, which I can then be taken as asserting that the action violates, but it is not bound to any such presupposition. It would not be invalidated if my moral sentiments were at variance with those that prevailed in my community or indeed in any other. I could be argued out of them, in this or another instance, not only by its being shown to me that I was not properly or fully informed of the facts, but in a variety of other ways. For instance, I may be convinced on philosophical grounds that law-breakers should not be regarded otherwise than as sufferers from disease; I may attach importance to maintaining consistency in my moral attitudes, and may be persuaded that this case does not differ significantly from others of which I have taken a different view: I may be given cause to think that my view of the case is clouded by untoward features of my own experience and character. I may consequently decide that my original moral judgement was mistaken. I say 'mistaken' rather than 'false', because I think it conducive to clarity to assign a truth-value only to moral judgements which carry the presupposition of some code, so that it becomes a factual question whether the judgement accords with the rules of measurement which the code furnishes. Evidently this does not imply either that any one code is sacrosanct, or that all moral judgements are equally acceptable.

Hume's contention that 'ought' does not follow from 'is' has also been challenged. The favourite counter-example is that of promising. It has been argued that from the purely factual premise that, under specifiable conditions, a man utters a sentence of the form 'I promise to do X', it logically follows that, other things being equal, he ought to do X. But this is a fallacy. If the argument seems convincing, it is because it is situated within a moral climate in which provision is made for promising, that is, for the incurring of a moral obligation through the utterance, under the proper conditions, of certain sequences of words. But if we are dealing with a question of logical entailment, the existence of such a climate cannot legitimately be presupposed. It has to be stated as an extra premise, to the effect that the speaker belongs to a society in which it is an accepted principle that to utter such and such words is, in some range of circumstances, to assume a moral commitment. This is again a factual premise, but even when conjoined with the other, it does not entail the desired conclusion. We still need the moral premise that this rule of his society is one to which allegiance ought to be given.

A simple objection to Hume's theory is that it applies equally well to natural endowments or handicaps, such as good or ill looks, wit or dullness, a sociable or morose disposition, as to what are commonly reckoned as virtues or vices, like courage or cowardice, fidelity or fickleness, meanness or generosity. In an Appendix to the *Enquiry,* Hume acknowledges the fact and dismisses it as of no importance, appealing on this point to the authority of Cicero, whom he frequently takes as a model, and also to that of Aristotle, who 'ranks courage, temperance, magnificence, modesty, prudence and a manly openness, among the virtues, as well as justice and friendship' (E 319). This fails to account for the distinction that we ordinarily make, though I suspect that Hume is right in claiming that it is not based on any consistent principle. Possibly, we tend to confine moral epithets to those amiable or unamiable qualities which require more cultivation to become habitual, or depend to a greater extent upon the presence or absence of self-discipline. This would still allow good manners to count among the virtues, but I do not find this unacceptable.

To anyone who has studied Kant, who notoriously held that an action is moral only when it is done from a sense of duty, it may come as a shock to find Hume saying that an action must proceed from some motive other than a sense of its morality if it is to be morally good. Hume does not deny that men can and do act out of a sense of duty. What he denies is that this in itself confers any merit on an action. A man who is of a miserly disposition may grow ashamed of it and so force himself to perform acts of generosity. In time his initial reluctance to perform them may or may not be overcome. It is not, however, necessary that it should be overcome in order for his actions to be morally good. Their goodness depends on their conforming to a habitual practice of generosity, and so long as this is true it makes no moral difference whether the agent has generous feelings, whether he thinks it profitable to himself to display a generosity which goes against his inclinations, or whether he acts against his inclinations because he thinks that generosity ought to be practised. We must, therefore, avoid being misled by Hume's saying that actions are morally good only as being signs of a good motive or a good character. It does not mean that he conceives of characters or motives as being good in themselves. They owe their goodness only to the fact that they *habitually* give rise to actions which are morally approved. It is the consequences that call the tune: and motives are brought to the fore only because they can be counted upon regularly to produce beneficent actions. To have a sense of duty as one's primary motive for behaving well is rather to be deprecated, since this suggests that one is deficient in natural benevolence.

In fact the opposition between Hume and Kant on this issue goes deeper, since Kant's ground for tying morality to the sense of duty was that an action could have moral worth only if it was performed freely, and that it was only when they were done from a sense of duty that actions were free in the requisite form. It is not at all clear what this form of freedom was supposed to be but in any case there seems to be no warrant for drawing a distinction on this basis between the sense of duty and any other motive.

The principle that reason alone can never be the motive for any action, though it may appear psychological, is defended by Hume in a way that makes it analytic. It is a straightforward consequence of his confining what he counts as reason to the drawing of inferences and the appraisal of truth and falsehood, and his semantic assumption that operative motives cannot significantly be characterised as true or false. Reason has control over the passions, in so far as it can be used to discover that a passion is based on a false judgement, as for example when the object of one's fear is proved not to exist, or that the means chosen to procure some end are insufficient for the purpose. A third instance, which Hume failed to notice, is that in which reason shows us that the achievement of a desired end will prob-ably result in the occurrence of something which we have a greater desire to avoid. The dramatic statement that reason is and ought only to be the slave of the passions amounts only to what is, for Hume, the truism that reason enters into the sphere of action only when we have been motivated to aim at some end. The phrase 'ought only to be' is merely a rhetorical flourish, since the choice of ends has been put outside the sphere of reason. Similarly, Hume's startling assertion that it is 'not contrary to reason to prefer the destruction of the whole world to the scratch-ing of my finger' or 'to choose my total ruin, to prevent the least uneasiness of an *Indian* or person wholly unknown to me' or 'to prefer even my own acknowledg'd lesser good to my greater' (T 416) is no more than a consequence of the scarcely con-troversial assumption that one's preferences, however eccentric, are not themselves the bearers of truth-values and need not be the outcome of any erroneous inferences.

The supporting argument that morals cannot be derived from reason, since they have an influence on our actions and affections is, on the other hand, invalid. Even on Hume's own showing, our actions and affections may be influenced by the truth or false-hood of our judgements or the soundness of our inferences; it is only our objectives, when not considered as means to further ends, that are not susceptible to this influence. Nevertheless the deduction that 'the rules of morality are not conclusions of our reason' holds good independently, since what it rightly claims is that the rules of morality do not come within the province of 'Relations of Ideas'.

The main objection to be faced is that we do have a conception of the rationality of ends as well as means. Hume avoids it by suggesting that we mistake calm passions for 'the determinations of reason' but this is not a sufficient answer. It is true that we speak of a person as behaving irrationally when he rushes into a course of action without thinking about the consequences, but we also consider some rankings of ends, such as those that occur in Hume's examples, as being irrational in their own right. I think that what we mean by this is that they are choices which no sensible man would make. There is an obvious risk of circularity here, since his choice of ends enters into our conception of what constitutes a sensible man. A possible way of escape is just to take it as a matter of fact that a person who habitually chooses ends of a certain type arouses in the average spectator an impression of folly. We can then define the irrationality of ends in terms of the behaviour of such a person, and their rationality as its opposite.

There remains the question whether the sense of justice, in which Hume discovers the source of our obligations, is natural or artificial. Hume's ground for saying that it is artificial is that there are no natural motives to supply it. With very rare excep-tions, men do not have any such passion as the love of mankind, and their natural feelings of benevolence towards a limited number of persons would favour injustice, in that it would lead them to promote the interest of these persons at the public cost. Neither can the sense of justice arise directly from man's selfishness, though Hume's explanation of it does relate it in-directly to self-interest.

His explanation runs as follows. Because of his physical weakness, a man can survive and prosper only as a member of a group, even if in the first instance it is so small a group as that of the family. As groups intermingle, they tend to prey on one another. 'There are three different species of goods, which we are possess'd of; the internal satisfactions of our mind, the external advantages of our body, and the enjoyment of such possessions as we have acquir'd by our industry and good fortune' (T 487). It is the third of these goods that is chiefly at risk, on the one hand because of men's avidity and the limited extent of their benevolence, and on the other because of the niggardliness of nature. If men had always lived in an environment of such abundance that all their material appetites, however luxurious, could easily be satisfied, 'the cautious jealous virtue of justice would never once have been dreamed of' (E 184). As things are, different groups have to compete for the relatively scarce quantity of goods that are or can be made available. If this competition were unrestrained, no man could ever count on 'the peaceable enjoyment of what he may acquire by his fortune and industry' (T 489). To avoid such a calamity men have found it in their interest to subscribe to a set of conventions which establish rights to property and also lay down the conditions in which its transfer from one man to another is legitimate. These conventions are not promises. On the contrary, promises are themselves based on a convention, the purpose of which is to give men security of each other's future conduct. It is on the three 'laws' of the stability of possession, its transference by consent, and the performance of promises 'that the peace and security of human society entirely depend' (T 526).

So far it is only a question of self-interest. Morality comes into the picture both naturally, because we sympathise with the victims of injustice, either through our affection for them, or, in cases which this does not cover, through putting ourselves imaginatively in their place, and artificially, because those who educate and govern us see it as their business to train us to apply eulogistic epithets to conformity with the rules of justice and dyslogistic epithets to their breach, thereby arousing and fortifying the moral sentiments with which the use of these epithets is associated.

The justification of these rules is that their observance conduces to the public interest. This is, however, true only in general and may be false in some particular instances. An example which Hume gives is that in which 'a man of a beneficent disposition, restores a great fortune to a miser or a seditious bigot'; the man 'has acted justly and laudably, but the public is a real sufferer' (T 497). Even so, Hume maintains, in agreement with many present-day utilitarians, that we should always stick to the general rule. He offers no argument for this beyond saying that it is 'impossible to separate the good from the ill. Property must be stable, and must be fix'd by general rules. Tho' in one instance the public be a sufferer, this momentary ill is amply compensated by the steady prosecution of the rule, and by the peace and order, which it establishes in society' (T 497). But why should the public be required to suffer at all? Only, it would seem, because toleration of exceptions would weaken respect for the general rule, with a resulting disutility that would outweigh the utility of the exceptions. But even if we accept this line of reasoning, it does not cover the cases where there is no serious probability that the choice of an action which went counter to the rule would become generally known. Why should we then adhere to the rule when the purpose of the rule would be better served by our departing from it? I do not believe that those who uphold what is currently known as rule-utilitarianism have a satisfactory answer to this question.

Another objection to Hume's account of justice is that it is tied too closely to the institution of property. For instance, the only provision that it makes for the value of equality is that we are equally bound to comply with the rules that our conventions establish. This is, however, a defect that is easily remedied, since the conventions can always be adjusted to suit any ranking of their interests that the members of a society can be induced to accept.

Though justice is necessary for the maintenance of society, government, in Hume's view, is not. A primitive uncultivated society can subsist without it. Only, when it is successful in war, a period of government will be needed to regulate the partition of the spoils. The warrior chief is then likely to become the civil arbiter. It is for this reason, according to Hume, that 'all governments are at first monarchical'.

In more developed societies, government is needful because men are naturally disposed to sacrifice their long-term to their short-term objectives, and it is in their interest to be restrained from doing so. It therefore pays them to submit to rulers, whose short-term interests include the retention of their privileges, which in turn depend upon the enforcement of law and order. There is, indeed, the possibility of the rulers' becoming so tyrannical that their subjects would be better off without them, in which case they are at liberty to remove them if they have the power to do so. Such action is not, however, to be undertaken lightly, in view of the disturbance which it creates, and it is also true that the habit of civil obedience leads men to put up with tyranny for longer than would be the case if they strictly attended to their interest. There are various ways, including succession, conquest, and the operation of constitutional laws, in which particular governments may be instituted. What is more important is the tenure of power. No matter how a system of government comes into being, its mere persistence will generally be sufficient for its being regarded as legitimate. There is a sense, then, in which government rests on the consent of the governed, but it does not follow from this that we have to invent a social contract as the moral justification for political obedience. Since the rules of justice are necessary for the validity of contracts and also sufficient to account for political obligation, even a genuine contract would, as Hume rightly argues, have no part to play in this case, let alone the philosophical fictions which writers on politics have tried to pass off as contracts. The suggestion has been made that we incur an obligation to obey a government by choosing to remain subject to it, but apart from any other objection this presupposes that we have a serious alternative, which is very often not the case. Otherwise, as Hume remarks in his essay 'Of the Original Contract', it is like telling a man who has been shanghaied aboard a ship that he is free to leap into the sea and perish.

Hume makes no attempt to connect morals with religion, no doubt because he saw that morals cannot be grounded on any form of authority, however powerful, though religious belief may operate as a sanction through its effect on the passions. In any case Hume was less interested in the utility of religious belief than in its pretension to truth. We have seen that he rejected Christianity. By examining his *Dialogues Concerning Natural Religion*, I shall now try to justify my claim that he also rejected deism.

The three participants in the *Dialogues* are Demea, who believes that the existence of God can be demonstrated *a priori*, though we are incapable of penetrating the mystery of his nature; Cleanthes, who argues that there can be no other proof of God's existence than what we can extract from our observations of the world, and consequently bases his theism upon what is commonly called the argument from design; and the sceptical Philo, who agrees with Cleanthes that the argument from design is the only one worth considering, and devotes himself to showing what a very poor argument it is. This does not lead him into any profession of atheism. He even goes so far at one point as to assert that 'when reasonable men treat these subjects, the question can never be concerning the *being*, but only the *nature* of the Deity' (D 142), but in the light of his general argument this remark appears ironical. In the management of the *Dialogues*, Philo is given by far the most to say, and this is one reason for concluding that he speaks for Hume.

Demea's argument is that nothing can exist without a cause, that the idea of an infinite regress of causes is absurd, and that the regress can be brought to an end only by there being an ultimate cause who necessarily exists. This is the Deity, 'who carries the *Reason* of existence in himself; and who cannot be supposed not to exist without an express contradiction' (D 189). Cleanthes's rebuttal of this argument is that the existence of a deity is supposed to be a matter of fact, and that no matter of fact is demonstrable *a priori*. Moreover, if we were entitled to speak of a necessary being, it might just as well be the universe itself. Demea makes no attempt to reply to this objection, with which Philo concurs, and his part in the remainder of the *Dialogues* is pretty well limited to an occasional pious protest against Cleanthes's failure to make a sufficiently large distinction between the qualities and powers of human beings and those of their supposed creator.

While he gives numerous examples, accepted by Philo, of the adaptation of means to ends in nature, Cleanthes's position is summarised in one powerful statement of the argument from design.

Look round the world [he says]: Contemplate the whole and every part of it: You will find it to be nothing but one great machine, subdivided into an infinite number of lesser machines, which again admit of subdivisions, to a degree beyond what human senses and faculties can trace and explain. All these various machines, and even their most minute parts, are adjusted to each other with an accuracy, which ravishes into admiration all men, who have ever contemplated them. The curious adapting of means to ends, throughout all nature, resembles exactly, though it much exceeds, the productions of human contrivance; of human design, thought, wisdom, and intelligence. Since therefore the effects resemble each other, we are led to infer, by all the rules of analogy, that the causes also resemble; and that the Author of nature is somewhat similar to the mind of man: though possessed of much larger faculties, proportioned to the grandeur of the work, which he has executed. By this argument *a posteriori*, and by this argument alone, we do prove at once the existence of a Deity, and his similarity to human mind and intelligence. (D 143)

Philo's rejoinder to this argument is the principle theme of the Dialogues as a whole, but his main objections to it can be concisely summarised.

1 Causal arguments are based on experienced regularities. They are not available in this case since we are not acquainted with a multiplicity of worlds. Cleanthes can appeal only to analogy, which is a weaker form of reasoning,

2 If we press the analogy, on the assumption that like effects have like causes, we have no warrant for concluding that the universe was planned by an infinite, eternal, incorporeal Being. We have no experience of anything of this sort. Machines are constructed by mortal human beings who have bodies, belong to one or other sex, work in co-operation, proceed by trial and error, make blunders and correct them, improve on their designs. By what right therefore can we deprive the universal planner of body and sex? Why should we not conclude that the world is due to the combined efforts of many gods? Why should it not be 'the first rude essay of some infant deity, who afterwards abandoned it' or 'the production of old age and dotage in some super-annuated Deity' (D 169) whose death left it to its own devices?

Why should not many worlds 'have been botched and bungled, throughout an eternity, ere this system was struck out' (D 167)?

3 If this is to make the analogy too strong to suit Cleanthes, it is also in fact too weak. The world does contain many human artefacts, and many natural objects which resemble these artefacts at least to the extent that they or their parts serve some function, to which they are more or less well adapted. It does not follow, however, that the universe as a whole is a machine or anything like one; or that there is any purpose which it serves. It is no more like a machine than it is like an animal or vegetable organism. It would be no 'less intelligent, or less conformable to experience to say, that the world arose from vegetation from a seed shed by another world, than to say that it arose from a divine reason or contrivance' (D 178).

4 If we are tracing the universe of objects into 'a similar universe of ideas' (D 162), why should we stop there? If the order in ideas needs no further explanation, why should the order that we find in matter?

5 Our experience of the world shows that 'matter can preserve that perpetual agitation, which seems essential to it, and yet maintain a constancy in the forms, which it produces' (D 183). Why should we not be content to credit matter with a force which enables it to develop out of what we may suppose to have been an original chaos into an order accounting among other things 'for all the appearing wisdom and contrivance which is in the universe' (D 184)? Surely this theory is to be preferred to the unverifiable and practically useless hypothesis of a supernatural agency.

6 Not only is the argument from design exposed to all these objections, but even if they are waived it achieves next to nothing. The most that a man who accepts it is entitled to conclude is 'that the universe, sometime, arose from something like design: But beyond this position he cannot ascertain one single circumstance, and is left afterwards to fix every point of his theology, by the utmost licence of fancy and hypothesis' (D 169).

Probably the last point was, in Hume's eyes, the most important of all. He was, as I have tried to show, campaigning on many fronts against religious belief, but above all he wished to preserve philosophy from the 'licence of fancy and hypothesis' into which theology falls. We have seen that he was not a model of consistency, but he was at least consistent in his naturalism, his insistence that every branch of science be anchored in experience. His main interests were too broad to be captured in a paragraph, but his general outlook can perhaps be summarised, and I can see no better way of doing so than by quoting the famous passage with which he concludes the first *Enquiry*:

When we run over libraries, persuaded of these principles, what havoc must we make? If we take in our hand any volume; of divinity or school metaphysics, for instance; let us ask, *Does it contain any abstract reasoning concerning quantity or number?* No. *Does it contain any experimental reasoning concerning matter of fact and existence?* No. Commit it then to the flames: for it can contain nothing but sophistry and illusion. (E 165)

Bibliographical note

The editions of Hume's works from which I have quoted are listed in the Preface. The original dates of publication of these, and of other works by Hume to which I refer, are as follows:

A Treatise of Human Nature 1739–40
An Abstract of the Treatise of Human Nature 1740 (published anonymously)
Essays, Moral and Political 1741–2
Three Essays ('Of Natural Character', 'Of the Original Contract' and 'Of Passive Obedience') 1748
Enquiry concerning Human Understanding 1748 (first published as *Philosophical Essays concerning Human Understanding*)
Enquiry concerning the Principles of Morals 1751
Political Discourses 1752
History of Great Britain from the Invasion of Julius Caesar to the Revolution of 1688 (6 vols) 1754–62
Four Dissertations ('The Natural History of Religion', 'Of the Passions', 'Of Tragedy' and 'Of the Standard of Taste') 1757
Two Essays ('Of Suicide' and 'Of the Immortality of the Soul') 1777
My Own Life 1777 (first published as *The Life of David Hume, Esq., Written by Himself*)
Dialogues concerning Natural Religion 1779

A large collection of Hume's letters was edited by J. Y. T. Greig in two volumes and published in 1932 under the title *The Letters of David Hume* by the Oxford University Press. In 1954 the same publishers brought out a collection entitled *New Letters of David Hume*, the product of extensive research by its editors, Raymond Klibansky and Ernest C. Mossner. Of the many books written about Hume's philosophy, the two which I should especially wish to recommend are *The Philosophy of David Hume* by Norman Kemp Smith, which was published by Macmillan in

1941, and *Hume's Theory of The External World* by H. H. Price, which was published by the Oxford University Press in 1940. The points for which I am indebted to these works are indicated in the text.

Index

Past Masters

AQUINAS Anthony Kenny

Anthony Kenny writes about Thomas Aquinas as a philosopher, for readers who may not share Aquinas's theological interests and beliefs. He begins with an account of Aquinas's life and works, and assesses his importance for contemporary philosophy. The book is completed by more detailed examinations of Aquinas's metaphysical system and his philosophy of mind.

DANTE George Holmes

George Holmes expresses Dante's powerful originality by identifying the unexpected connections the poet made between received ideas and his own experience. He presents Dante's biography both as an expression of the intellectual dilemma of early Renaissance Florence and as an explanation of the poetic, philosophical and religious themes developed in his works. He ends with a discussion of the *Divine Comedy*, Dante's poetic panorama of hell, purgatory and heaven.

JESUS Humphrey Carpenter

Humphrey Carpenter writes about Jesus from the standpoint of a historian coming fresh to the subject without religious preconceptions. He examines the reliability of the Gospels, the originality of Jesus's teaching, and Jesus's view of himself. His highly readable book achieves a remarkable degree of objectivity about a subject which is deeply embedded in Western culture.

Past Masters

MARX Peter Singer

Peter Singer identifies the central vision that unifies Marx's thought, enabling us to grasp Marx's views as a whole. He views him as a philosopher primarily concerned with human freedom, rather than as an economist or social scientist. He explains alienation, historical materialism, the economic theory of *Capital*, and Marx's idea of communism, in plain English, and concludes with a balanced assessment of Marx's achievement.

PASCAL Alban Krailsheimer

Alban Krailsheimer opens his study of Pascal's life and work with a description of Pascal's religious conversion, and then discusses his literary, mathematical and scientific achievements, which culminated in the acute analysis of human character and powerful reasoning of the *Pensées*. He argues that after his conversion Pascal put his previous work in a different perspective and saw his, and in general all human activity, in religious terms.